CALCULATING
THE COSMOS

By the same author

Concepts of Modern Mathematics
Game, Set and Math
Does God Play Dice?
Another Fine Math You've Got Me Into ...
Fearful Symmetry (with Martin Golubitsky)
Nature's Numbers
From Here to Infinity
The Magical Maze
Life's Other Secret
Flatterland
What Shape is a Snowflake?
The Annotated Flatland
Math Hysteria
The Mayor of Uglyville's Dilemma
How to Cut a Cake
Letters to a Young Mathematician
Taming the Infinite (Alternative Title: The Story Of Mathematics)
Why Beauty is Truth
Cows in the Maze
Professor Stewart's Cabinet of Mathematical Curiosities
Mathematics of Life
Professor Stewart's Hoard of Mathematical Treasures
17 Equations that Changed the World (Alternative Title: In Pursuit of the Unknown)
The Great Mathematical Problems (Alternative Title: Visions of Infinity)
Symmetry: A Very Short Introduction
Jack of All Trades (Science Fiction eBook)
Professor Stewart's Casebook of Mathematical Mysteries
Professor Stewart's Incredible Numbers

with Jack Cohen
The Collapse Of Chaos
Evolving the Alien (Alternative Title: What Does A Martian Look Like?)
Figments of Reality
Wheelers (Science Fiction)
Heaven (Science Fiction)

The Science of Discworld Series (With Terry Pratchett And Jack Cohen)
The Science of Discworld
The Science of Discworld II: The Globe
The Science of Discworld III: Darwin's Watch
The Science of Discworld IV: Judgement Day

iPad app
Incredible Numbers

CALCULATING THE COSMOS

How MATHEMATICS UNVEILS *the* UNIVERSE

IAN STEWART

BASIC BOOKS

New York

Basic Books
Hachette Book Group
1290 Avenue of the Americas, New York, NY 10104
www.basicbooks.com/

Printed in the United States of America

First published in Great Britain in 2016 by Profile Books Ltd
3 Holford Yard, Bevin Way, London WC1X 9HD
www.profilebooks.com

First US Trade Paperback Edition: May 2018

Published by Basic Books, an imprint of Perseus Books, LLC, a subsidiary of Hachette Book Group, Inc. The Basic Books name and logo is a trademark of the Hachette Book Group.

The Hachette Speakers Bureau provides a wide range of authors for speaking events. To find out more, go to www.hachettespeakersbureau.com or call (866) 376-6591.

The publisher is not responsible for websites (or their content) that are not owned by the publisher.

Typeset in Sabon by MacGuru Ltd

A catalog record for this book is available from the Library of Congress.

LCCN: 2016935674
ISBN: 978-0-465-09610-7 (hardcover)
ISBN: 978-0-465-09611-4 (e-book)
ISBN: 978-1-5416-1725-4 (paperback)

LSC-C

10 9 8 7 6 5 4 3 2 1

Contents

Prologue / 1

1 Attraction at a Distance / 11

2 Collapse of the Solar Nebula / 27

3 Inconstant Moon / 40

4 The Clockwork Cosmos / 54

5 Celestial Police / 70

6 The Planet that Swallowed its Children / 84

7 Cosimo's Stars / 96

8 Off on a Comet / 106

9 Chaos in the Cosmos / 119

10 The Interplanetary Superhighway / 137

11 Great Balls of Fire / 150

12 Great Sky River / 172

13 Alien Worlds / 187

14 Dark Stars / 207

15 Skeins and Voids / 226

16 The Cosmic Egg / 241

17 The Big Blow-Up / 251

18 The Dark Side / 262

19 Outside the Universe / 277

Epilogue / 295

Units and Jargon / 299

Notes and References / 304

Picture Credits / 321

Index / 323

Prologue

'Why, I have calculated it.'

Isaac Newton's reply to Edmond Halley, when asked how he knew that an inverse square law of attraction implies that the orbit of a planet is an ellipse. Quoted in Herbert Westren Turnbull, *The Great Mathematicians*.

ON 12 NOVEMBER 2014 an intelligent alien observing the solar system would have witnessed a puzzling event. For months, a tiny machine had been following a comet along its path round the Sun – passive, dormant. Suddenly, the machine awoke, and spat out an even tinier machine. This descended towards the coal-black surface of the comet, hit ... and bounced. When it finally came to a halt, it was tipped over on its side and jammed against a cliff.

The alien, deducing that the landing had not gone as intended, might not have been terribly impressed, but the engineers behind the two machines had pulled off an unprecedented feat – landing a space probe on a comet. The larger machine was *Rosetta*, the smaller *Philae*, and the comet was 67P/Churyumov–Gerasimenko. The mission was carried out by the European Space Agency, and the flight alone took more than ten years. Despite the bumpy landing, *Philae* attained most of its scientific objectives and sent back vital data. *Rosetta* continues to perform as planned.

Why land on a comet? Comets are intriguing in their own right, and anything we can discover about them is a useful addition to basic science. On a more practical level, comets occasionally come close to the Earth, and a collision would cause huge devastation, so it's prudent to find out what they're made of. You can change the orbit of a solid body using a rocket or a nuclear missile, but a soft spongy one might break up and make the problem worse. However, there's a third reason. Comets contain material that goes back to the origin of the solar system, so they can provide useful clues about how our world came into being.

Astronomers think that comets are dirty snowballs, ice covered in a thin layer of dust. *Philae* managed to confirm this, for comet 67P, before its batteries discharged and it went silent. If the Earth formed at its current distance from the Sun, it has more water than it ought to. Where did the extra water come from? An attractive possibility is bombardment by millions of comets when the solar system was forming. The ice melted, and the oceans were born. Perhaps surprisingly, there's a way to test this theory. Water is made from hydrogen and oxygen. Hydrogen occurs in three distinct atomic forms, known as isotopes; these all have the same number of protons and electrons (one of each), but differ in the number of neutrons. Ordinary hydrogen has no neutrons, deuterium has one, and tritium has two. If Earth's oceans came from comets, the proportions of those isotopes in the oceans, and in the crust, whose rocks also contain large amounts of water within their chemical make-up, should be similar to their proportions in comets.

Philae's analysis shows that compared to Earth, 67P has a much greater proportion of deuterium. More data from other comets will be needed to make sure, but a cometary origin for the oceans is starting to look shaky. Asteroids are a better bet.

The 'rubber duck' comet 67P, imaged by Rosetta.

✦

The *Rosetta* mission is just one example of humanity's growing ability to send machines into space, either for scientific exploration or for everyday use. This new technology has expanded our scientific aspirations. Our space probes have now visited, and sent back their holiday snaps of, every planet in the solar system, and some of the smaller bodies.

Progress has been rapid. American astronauts landed on the Moon in 1969. The *Pioneer 10* spacecraft, launched in 1972, visited Jupiter and continued out of the solar system. *Pioneer 11* followed it in 1973 and visited Saturn as well. In 1977 *Voyager 1* and *Voyager 2* set out to explore these worlds, and the even more distant planets Uranus and Neptune. Other craft, launched by several different nations or national groupings, have visited Mercury, Venus, and Mars. Some have even *landed* on Venus and Mars, sending back valuable information. As I write in 2015, five orbital probes[1] and two[2] surface vehicles are exploring Mars, *Cassini* is in orbit around Saturn, the *Dawn* spacecraft is orbiting the former asteroid and recently promoted dwarf planet Ceres, and the *New Horizons* spacecraft has just whizzed past, and sent back stunning images of, the most famous dwarf planet in the solar system: Pluto. Its data will help resolve the mysteries of this enigmatic body and its five moons. It has already shown that Pluto is marginally larger than Eris, a more distant dwarf planet previously thought to be the largest. Pluto was reclassified as a dwarf planet in order to exclude Eris from planetary status. Now we discover they needn't have bothered.

We're also starting to explore lesser but equally fascinating bodies: moons, asteroids, and comets. It may not be *Star Trek*, but the final frontier is opening up.

Space exploration is basic science, and while most of us are intrigued by new discoveries about the planets, some prefer their tax contributions to produce more down-to-earth payoffs. As far as everyday life is concerned, our ability to create accurate mathematical models of bodies interacting under gravity has given the world a range of technological wonders that rely on artificial satellites: satellite television, a highly efficient international telephone network, weather satellites, satellites watching the Sun for magnetic storms, satellites keeping

On 14 July 2015 NASA's New Horizons space probe sent this historic image of Pluto to Earth, the first to show clear features on the dwarf planet.

watch on the environment and mapping the globe – even car satnav, using the Global Positioning System.

These accomplishments would have amazed previous generations. Even in the 1930s, most people thought that no human would ever stand on the Moon. (Today plenty of rather naive conspiracy theorists still think nobody has, but don't get me started.) There were heated arguments about even the bare *possibility* of spaceflight.[3] Some people insisted that rockets wouldn't work in space because 'there is nothing there to push against', unaware of Newton's third law of motion – to every action there is an equal and opposite reaction.[4]

Serious scientists stoutly insisted that a rocket would never work because you needed a lot of fuel to lift the rocket, then more fuel to lift the fuel, then more fuel to lift *that*… even when a picture in the fourteenth-century Chinese *Huolongjing* (Fire Dragon Manual) by Jiao Yu depicts a fire-dragon, aka multistage rocket. This Chinese naval weapon used discardable boosters to launch an upper stage shaped like the head of a dragon, loaded with fire arrows that shot out of its mouth. Conrad Haas made the first European experiment with multistage rockets in 1551. Twentieth-century rocketry pioneers pointed out that the first stage of a multistage rocket would be able to lift the second stage and its fuel, while *dropping off* all of the excess weight of the now-exhausted first stage. Konstantin Tsiolkovsky published detailed and realistic calculations about the exploration of the solar system in 1911.

Well, we got to the Moon despite the naysayers – using precisely the ideas that they were too blinkered to contemplate. So far, we've explored only our local region of space, which pales into insignificance compared to the vast reaches of the universe. We've not yet landed humans on another planet, and even the nearest star seems utterly out of reach. With existing technology, it would take centuries to get there, even if we could build a reliable starship. But we're on our way.

✦

These advances in space exploration and usage depend not just on clever technology, but also on a lengthy series of scientific discoveries that go back at least as far as ancient Babylon three millennia ago. Mathematics lies at the heart of these advances. Engineering is of course vital too, and discoveries in many other scientific disciplines were needed before we could make the necessary materials and assemble them into a working space probe, but I'll concentrate on how mathematics has improved our knowledge of the universe.

The story of space exploration and the story of mathematics have gone hand in hand from the earliest times. Mathematics has proved essential for understanding the Sun, Moon, planets, stars, and the vast panoply of associated objects that together form the cosmos – the universe considered on a grand scale. For thousands of years, mathematics has been our most effective method of understanding, recording, and predicting cosmic events. Indeed in some cultures, such as ancient India around 500, mathematics was a sub-branch of astronomy. Conversely, astronomical phenomena have influenced the development of mathematics for over three millennia, inspiring everything from Babylonian predictions of eclipses to calculus, chaos, and the curvature of spacetime.

Initially, the main astronomical role of mathematics was to record observations and perform useful calculations about phenomena such as solar eclipses, where the Moon temporarily obscures the Sun, or lunar eclipses, where the Earth's shadow obscures the Moon. By thinking about the geometry of the solar system, astronomical pioneers realised that the Earth goes round the Sun, even though it looks the other way round from down here. The ancients also combined observations with geometry to estimate the size of the Earth and the distances to the Moon and the Sun.

Deeper astronomical patterns began to emerge around 1600, when Johannes Kepler discovered three mathematical regularities – 'laws' – in the orbits of the planets. In 1679 Isaac Newton reinterpreted Kepler's laws to formulate an ambitious theory that described not just how the planets of the solar system move, but the motion of *any* system of celestial bodies. This was his theory of gravity, one of the central discoveries in his world-changing *Philosophiae Naturalis Principia Mathematica* (Mathematical Principles of Natural Philosophy). Newton's law of gravity describes how each body in the universe attracts every other body.

By combining gravity with other mathematical laws about the motion of bodies, pioneered by Galileo a century earlier, Newton explained and predicted numerous celestial phenomena. More generally, he changed how we think about the natural world, creating a scientific revolution that is still powering ahead today. Newton showed that natural phenomena are (often) governed by mathematical patterns, and by understanding these patterns we can improve our understanding of nature. In Newton's era the mathematical laws explained what was happening in the heavens, but they had no significant practical uses, other than for navigation.

✦

All that changed when the USSR's *Sputnik* satellite went into low Earth orbit in 1957, firing the starting gun for the space race. If you watch football on satellite television – or opera or comedies or science documentaries – you're reaping a real-world benefit from Newton's insights.

Initially, his successes led to a view of the cosmos as a clockwork universe, in which everything majestically follows paths laid down at the dawn of creation. For example, it was believed that the solar system was created in pretty much its current state, with the same planets moving along the same near-circular orbits. Admittedly, everything jiggled around a bit; the period's advances in astronomical observations had made that abundantly clear. But there was a widespread belief that nothing had changed, did change, or would change in any dramatic manner over countless eons. In European religion it was unthinkable that God's perfect creation could have been different in the past. The mechanistic view of a regular, predictable cosmos persisted for three hundred years.

No longer. Recent innovations in mathematics, such as chaos theory, coupled to today's powerful computers, able to crunch the relevant numbers with unprecedented speed, have greatly changed our views of the cosmos. The clockwork model of the solar system remains valid over short periods of time, and in astronomy a million years is usually short. But our cosmic backyard is now revealed as a place where worlds did, and will, migrate from one orbit to another. Yes, there are very long periods of regular behaviour, but from time to time they are punctuated by bursts of wild activity. The immutable laws that gave rise to the notion of a clockwork universe can also cause sudden changes and highly erratic behaviour.

The scenarios that astronomers now envisage are often dramatic. During the formation of the solar system, for instance, entire worlds collided with apocalyptic consequences. One day, in the distant future, they will probably do so again: there's a small chance that either Mercury or Venus is doomed, but we don't know which. It could be both, and they could take us with them. One such collision probably led to the formation of the Moon. It sounds like something out of science fiction, and it is ... but the best kind, 'hard' science fiction in which only the fantastic new invention goes beyond known science. Except that here there is no fantastic invention, just an unexpected mathematical discovery.

Mathematics has informed our understanding of the cosmos on every scale: the origin and motion of the Moon, the movements and form of the planets and their companion moons, the intricacies of asteroids, comets, and Kuiper belt objects, and the ponderous celestial dance of the entire solar system. It has taught us how interactions with Jupiter can fling asteroids towards Mars, and thence the Earth; why Saturn is not alone in possessing rings; how its rings formed to begin with and why they behave as they do, with braids, ripples, and strange rotating 'spokes'. It has shown us how a planet's rings can spit out moons, one at a time.

Clockwork has given way to fireworks.

✦

From a cosmic viewpoint, the solar system is merely one insignificant bunch of rocks among quadrillions. When we contemplate the universe on a grander scale, mathematics plays an even more crucial role. Experiments are seldom possible and direct observations are difficult, so we have to make indirect inferences instead. People with an anti-science agenda often attack this feature as some kind of weakness. Actually, one of the great strengths of science is the ability to infer things that we can't observe directly from those that we can. The existence of atoms was conclusively established long before ingenious microscopes allowed us to see them, and even then 'seeing' depends on a series of inferences about how the images concerned are formed.

Mathematics is a powerful inference engine: it lets us deduce the *consequences* of alternative hypotheses by pursuing their logical implications. When coupled with nuclear physics – itself highly mathematical – it helps to explain the dynamics of stars, with their many types, their different chemical and nuclear constitutions, their writhing magnetic fields and dark sunspots. It provides insight into the tendency of stars to cluster into vast galaxies, separated by even vaster voids, and explains why galaxies have such interesting shapes. It tells us why galaxies combine to form galactic clusters, separated by even vaster voids.

There's a yet larger scale, that of the universe as a whole. This is the realm of cosmology. Here humanity's source of rational inspiration is almost entirely mathematical. We can observe some aspects of the universe, but we can't experiment on it as a whole. Mathematics helps us to interpret observations, by permitting 'what if' comparisons between alternative theories. But even here, the starting point was closer to home. Albert Einstein's general theory of relativity, in which the force of gravity is replaced by the curvature of spacetime, replaced Newtonian physics. The ancient geometers and philosophers would have approved: dynamics was reduced to geometry. Einstein saw his theories verified by two of his own predictions: known, but puzzling, changes to the orbit of Mercury, and the bending of light by the Sun, observed during a solar eclipse in 1919. But he couldn't have realised that his theory would lead to the discovery of some of the most bizarre objects in the entire universe: black holes, so massive that even light can't escape their gravitational pull.

He certainly failed to recognise one potential consequence of his theory, the Big Bang. This is the proposal that the universe originated from a single point at some time in the distant past, around 13·8 billion years ago according to current estimates, in a sort of gigantic explosion. But it was spacetime that exploded, not something else exploding within spacetime. The first evidence for this theory was Edwin Hubble's discovery that the universe is expanding. Run everything backwards and it all collapses to a point; now restart time in the normal direction to get back to here and now.

Einstein lamented that he could have predicted this, if he'd believed his own equations. That's why we can be confident that he didn't expect it.

In science, new answers open up new mysteries. One of the greatest is dark matter, a completely new kind of matter that seems to be required to reconcile observations of how galaxies spin with our understanding of gravity. However, searches for dark matter have consistently failed to detect any. Moreover, two other add-ons to the original Big Bang theory are also required to make sense of the cosmos. One is inflation, an effect that caused the early universe to grow by a truly enormous amount in a truly tiny instant of time. It's needed to explain why the distribution of matter in today's universe is almost, but not quite, uniform. The other is dark energy, a mysterious force that causes the universe to expand at an ever-faster rate.

The Big Bang is accepted by most cosmologists, but only when these three extras – dark matter, inflation, and dark energy – are thrown in to the mix. However, as we shall see, each of these *dei ex machina* comes with a host of troubling problems of its own. Modern cosmology no longer seems as secure as it was a decade ago, and there might be a revolution on the way.

✦

Newton's law of gravity wasn't the first mathematical pattern to be discerned in the heavens, but it crystallised the whole approach, as well as going far beyond anything that had come before. It's a core theme of *Calculating the Cosmos*, a key discovery that lies at the heart of the book. Namely: there are mathematical patterns in the motions and structure of both celestial and terrestrial bodies, from the smallest dust particle to the universe as a whole. Understanding those patterns allows us not just to explain the cosmos, but also to explore it, exploit it, and protect ourselves against it.

Arguably the greatest breakthrough is to realise that there *are* patterns. After that, you know what to look for, and while it may be difficult to pin the answers down, the problems become a matter of technique. Entirely new mathematical ideas often have to be invented – I'm not claiming it's easy or straightforward. It's a long-running game and it's still playing out.

Newton's approach also triggered a standard reflex. As soon as the latest discovery hatches from its shell, mathematicians start wondering whether a similar idea might solve other problems. The urge to make everything more general runs deep in the mathematical psyche. It's no good blaming it on Nicolas Bourbaki[5] and the 'new maths': it goes back to Euclid and Pythagoras. From this reflex, mathematical physics was born. Newton's contemporaries, mainly in continental Europe, applied the same principles that had plumbed the cosmos to understand heat, sound, light, elasticity, and later electricity and magnetism. And the message rang out ever clearer:

> *Nature has laws.*
> *They are mathematical.*
> *We can find them.*
> *We can use them.*

Of course, it wasn't that simple.

Attraction at a Distance

Macavity, Macavity, there's no one like Macavity,
He's broken every human law, he breaks the law of gravity.
Thomas Stearns Eliot, *Old Possum's Book of Practical Cats*

WHY DO THINGS FALL down?

Some don't. Macavity, obviously. Along with the Sun, the Moon, and almost everything else 'up there' in the heavens. Though rocks sometimes fall from the sky, as the dinosaurs discovered to their dismay. Down here, if you want to be picky, insects, birds, and bats fly, but they don't stay up indefinitely. Pretty much everything else falls, unless something is holding it up. But up there, nothing holds it up – yet it doesn't fall.

Up there seems very different from down here.

It took a stroke of genius to realise that what makes terrestrial objects fall is the same thing that holds celestial objects up. Newton famously compared a falling apple to the Moon, and realised that the Moon stays up because, unlike the apple, it's also moving *sideways*.[1] Actually, the Moon is perpetually falling, but the Earth's surface falls away from it at the same rate. So the Moon can fall forever, yet go round and round the Earth and never hit it.

The real difference was not that apples fall and Moons don't. It was that apples don't move sideways fast enough to miss the Earth.

Newton was a mathematician (and a physicist, chemist, and mystic), so he did some sums to confirm this radical idea. He calculated the forces that must be acting on the apple and the Moon to make them follow their separate paths. Taking their different masses into account, the forces turned out to be identical. This convinced him that the Earth must be pulling both apple and Moon towards it. It was natural to suppose that the same type of attraction holds for any pair of bodies, terrestrial or celestial. Newton expressed those attractive forces in a mathematical equation, a law of nature.

One remarkable consequence is that not only does the Earth attract the apple: the apple also attracts the Earth. And the Moon, and everything else in the universe. But the apple's effect on the Earth is way too small to measure, unlike the Earth's effect on the apple.

This discovery was a huge triumph, a deep and precise link between mathematics and the natural world. It also had another important implication, easily missed among the mathematical technicalities: despite appearances, 'up there' is in some vital respects the same as 'down here'. The laws are identical. What differs is the context in which they apply.

We call Newton's mysterious force 'gravity'. We can calculate its effects with exquisite accuracy. We still don't understand it.

✦

For a long time, we thought we did. Around 350 BC the Greek philosopher Aristotle gave a simple reason why objects fall down: they are seeking their natural resting place.

To avoid circular reasoning, he also explained what 'natural' meant. He maintained that everything is made from four basic elements: earth, water, air, and fire. The natural resting place of earth and water are at the centre of the universe, which of course coincides with the centre of the Earth. As proof, the Earth doesn't move: we live on it, and would surely notice if it did. Since earth is heavier than water (it sinks, right?) the lowest regions are occupied by earth, a sphere. Next comes a spherical shell of water, then one of air (air is lighter than water: bubbles rise). Above that – but lower than the celestial sphere that carries the Moon – is the realm of fire. All other bodies tend to rise or fall according to the proportions in which these four elements occur.

This theory led Aristotle to argue that the speed of a falling body is proportional to its weight (feathers fall more slowly than stones) and inversely proportional to the density of the surrounding medium (stones fall faster in air than in water). Having reached its natural rest state, the body remains there, moving only when a force is applied.

As theories go, these aren't so bad. In particular, they agree with everyday experience. On my desk, as I write, there is a first edition of the novel *Triplanetary*, quoted in the epigram for Chapter 2. If I leave it alone, it stays where it is. If I apply a force – give it a shove – it moves a few centimetres, slowing down as it does so, and stops.

Aristotle was right.

And so it seemed for nigh on two thousand years. Aristotelian physics, though widely debated, was generally accepted by almost all intellectuals until the end of the sixteenth century. An exception was the Arab scholar al-Hasan ibn al-Haytham (Alhazen), who argued against Aristotle's view on geometric grounds in the eleventh century. But even today, Aristotelian physics matches our intuition more closely than do the ideas of Galileo and Newton that replaced it.

To modern thinking, Aristotle's theory has some big gaps. One is weight. *Why* is a feather lighter than a stone? Another is friction. Suppose I placed my copy of *Triplanetary* on an ice-skating rink and gave it the same push. What would happen? It would go further: a lot further if I rested it on a pair of skates. Friction makes a body move more slowly in a viscous – sticky – medium. In everyday life, friction is everywhere, and that's why Aristotelian physics matches our intuition better than Galilean and Newtonian physics do. Our brains have evolved an internal model of motion with friction built in.

Now we know that a body falls towards the Earth because the planet's gravity pulls it. But what is gravity? Newton thought it was a force, but he didn't explain how the force arose. It just *was*. It acted at a distance without anything in between. He didn't explain how it did that either; it just *did*. Einstein replaced force by the curvature of spacetime, making 'action at a distance' irrelevant, and he wrote down equations for how curvature is affected by a distribution of matter – but he didn't explain *why* curvature behaves like that.

People calculated aspects of the cosmos, such as eclipses, for millennia before anyone realised that gravity existed. But once gravity's role was revealed, our ability to calculate the cosmos became far more powerful. Newton's subtitle for Book 3 of the *Principia*, which described his laws of motion and gravity, was 'Of the System of the World'. It was only a slight exaggeration. The force of gravity, and the manner in which bodies respond to forces, lie at the heart of most cosmic calculations. So before we get to the latest discoveries, such as how ringed planets spit out moons, or how the universe began, we'd better sort out some basic ideas about gravity.

✦

Before the invention of street lighting, the Moon and stars were as familiar, to most people, as rivers, trees, and mountains. As the Sun went down, the stars came out. The Moon marched to its own drummer, sometimes appearing during the day as a pale ghost, but shining much more brightly at night. Yet there were patterns. Anyone observing the Moon even casually for a few months would quickly notice that it follows a regular rhythm, changing shape from a thin crescent to a circular disc and back again every 28 days. It also moves noticeably from one night to the next, tracing a closed, repetitive path across the heavens.

The stars have their own rhythm too. They revolve, once a day, round a fixed point in the sky, as if they're painted on the inside of a slowly spinning bowl. *Genesis* talks of the firmament of Heaven: the Hebrew word translated as 'firmament' means bowl.

Observing the sky for a few months, it also became obvious that five stars, including some of the brightest, don't revolve like the majority of 'fixed' stars. Instead of being attached to the bowl, they crawl slowly across it. The Greeks associated these errant specks of light with Hermes (messenger of the gods), Aphrodite (goddess of love), Ares (god of war), Zeus (king of the gods), and Kronos (god of agriculture). The corresponding Roman deities gave them their current English names: Mercury, Venus, Mars, Jupiter, and Saturn. The Greeks called them *planetes*, 'wanderers', hence the modern name planets, of which we now recognise three more: Earth, Uranus, and Neptune. Their paths were strange, seemingly unpredictable. Some moved relatively quickly, others were slower. Some even looped back on themselves as the months passed.

Most people just accepted the lights for what they were, in the same way that they accepted the existence of rivers, trees, and mountains. But a few asked questions. What are these lights? Why are they there? How and why do they move? Why do some movements show patterns, while others break them?

The Sumerians and Babylonians provided basic observational data. They wrote on clay tablets in a script known as cuneiform – wedge-shaped. Among the Babylonian tablets that archaeologists have found are star catalogues, listing the positions of stars in the sky; they date to about 1200 BC but were probably copies of even earlier Sumerian tablets. The Greek philosophers and geometers who followed their lead were more aware of the need for logic, proof, and theory. They were pattern-seekers; the Pythagorean cult took this attitude to extremes, believing that the entire universe is run by numbers. Today most scientists would agree, but not about the details.

The Greek geometer who had the most influence on the astronomical thinking of later generations was Claudius Ptolemy, an astronomer and geographer. His earliest work is known as the *Almagest*, from an Arabic rendering of its original title, which started out as 'The Mathematical Compilation', morphed into 'The Great Compilation', and then into '*al-majisti*' – the greatest. The *Almagest* presented a fully fledged theory of planetary motion based on what the Greeks considered to be the most perfect of geometric forms, circles and spheres.

The planets do not, in fact, move in circles. This wouldn't have been news to the Babylonians, because it doesn't match their tables. The Greeks went further, asking what would match. Ptolemy's answer was: combinations of circles supported by spheres. The innermost sphere, the 'deferent', is centred on the Earth. The axis of the second sphere, or 'epicycle', is fixed to the sphere just inside it. Each pair of spheres is disconnected from the others. It wasn't a new idea. Two centuries earlier, Aristotle – building on even earlier ideas of the same kind – had proposed a complex system of 55 concentric spheres, with the axis of each sphere fixed to the sphere just inside it. Ptolemy's modification used fewer spheres, and was more accurate, but it was still rather complicated. Both led to the question whether the spheres actually existed, or were just convenient fictions – or whether something entirely different was really going on.

✦

For the next thousand years and more, Europe turned to matters theological and philosophical, basing most of its understanding of the natural world on what Aristotle had said around 350 BC. The universe was believed to be geocentric, with everything revolving around a stationary Earth. The torch of innovation in astronomy and mathematics passed to Arabia, India, and China. With the dawn of the Italian Renaissance, however, the torch passed back to Europe. Subsequently, three giants of science played leading roles in the advance of astronomical knowledge: Galileo, Kepler, and Newton. The supporting cast was huge.

Galileo is famous for his invention of improvements to the telescope, with which he discovered that the Sun has spots, Jupiter has (at least) four moons, Venus has phases like the Moon's, and there's something strange about Saturn – later explained as its ring system. This evidence led him to reject the geocentric theory and embrace Nicolaus Copernicus's rival heliocentric theory, in which the planets and the Earth revolve round the Sun, getting Galileo into trouble with the Church of Rome. But he also made an apparently more modest, but ultimately more important, discovery: a mathematical pattern in the motion of objects such as cannonballs. Down here, a freely moving body either speeds up (when falling) or slows down (when rising) by an amount that is the same over a fixed, *small* period of time. In short, the body's acceleration is constant. Lacking accurate clocks, Galileo observed these effects by rolling balls down gentle inclines.

The next key figure is Kepler. His boss Tycho Brahe had made very accurate measurements of the position of Mars. When Tycho died, Kepler inherited his position as astronomer to Holy Roman Emperor Rudolph II, together with his observations, and set about calculating the true shape of Mars's orbit. After fifty failures, he deduced that the orbit is shaped like an ellipse – an oval, like a squashed circle. The Sun lies at a special point, the focus of the ellipse.

Ellipses were familiar to the ancient Greek geometers, who defined them as plane sections of a cone. Depending on the angle of the plane relative to the cone, these 'conic sections' include circles, ellipses, parabolas, and hyperbolas.

When a planet moves in an ellipse, its distance from the Sun varies. When it comes close to the Sun, it speeds up; when it's more distant, it slows down. It's a bit of a surprise that these effects conspire to create an orbit that has exactly the same shape at both ends. Kepler didn't expect this, and for a long time it persuaded him that an ellipse must be the wrong answer.

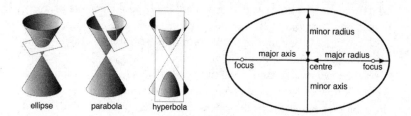

Left: Conic sections. Right: Basic features of an ellipse.

The shape and size of an ellipse are determined by two lengths: its major axis, which is the longest line between two points on the ellipse, and its minor axis, which is perpendicular to the major axis. A circle is a special type of ellipse for which these two distances are equal; they then give the diameter of the circle. For astronomical purposes the radius is a more natural measure – the radius of a circular orbit is the planet's distance from the Sun – and the corresponding quantities for an ellipse are called the major radius and minor radius. These are often referred to by the awkward terms semi-major axis and semi-minor axis, because they cut the axes in half. Less intuitive but very important is the eccentricity of the ellipse, which quantifies how long and thin it is. The eccentricity is 0 for a circle and for a fixed major radius it becomes infinitely large as the minor radius tends to zero.[2]

The size and shape of an elliptical orbit can be characterised by two numbers. The usual choice is the major radius and the eccentricity. The minor radius can be found from these. The Earth's orbit has major radius 149·6 million kilometres and eccentricity 0·0167. The minor radius is 149·58 million kilometres, so the orbit is very close to a circle, as the small eccentricity indicates. The plane of the Earth's orbit has a special name: the ecliptic.

The spatial location of any other elliptical orbit about the Sun can be characterised by three more numbers, all angles. One is the inclination of the orbital plane to the ecliptic. The second effectively gives the direction of the major axis in that plane. The third gives the direction of the line at which the two planes meet. Finally, we need to know where the planet is in the orbit, which requires one further angle. So specifying the orbit of the planet and its position within that orbit requires two numbers and four angles – six *orbital elements*. A major goal of early astronomy was to calculate the orbital elements of every planet and asteroid that was discovered. Given these numbers, you can predict its future motion, at least until the combined effects of the other bodies disturb its orbit significantly.

Kepler eventually came up with a set of three elegant mathematical patterns, now called his laws of planetary motion. The first states that the orbit of a planet is an ellipse with the Sun at one focus. The second says that the line from the Sun to the planet sweeps out equal areas in equal periods of time. And the third tells us that the square of the period of revolution is proportional to the cube of the distance.

✦

Newton reformulated Galileo's observations about freely moving bodies as three laws of motion. The first states that bodies continue to move in a straight line at a constant speed unless acted on by a force. The second states that the acceleration of any body, multiplied by its mass, is equal to the force acting on it. The third states that every action produces an equal and opposite reaction. In 1687 he reformulated Kepler's planetary laws as a general rule for how heavenly bodies move – the law of gravity, a mathematical formula for the gravitational force with which any body attracts any other.

Indeed, he *deduced* his force law from Kepler's laws by making one assumption: the Sun exerts an attractive force, always directed towards its centre. On this assumption, Newton proved that the force is inversely proportional to the square of the distance. That's a fancy way to say that, for example, multiplying the mass of either body by three also trebles the force, but multiplying the distance between them by three reduces the force to one ninth of the amount. Newton also proved the converse: this 'inverse square law' of attraction implies Kepler's three laws.

Credit for the law of gravity rightly goes to Newton, but the idea wasn't original with him. Kepler deduced something similar by analogy with light, but thought gravity pushed planets round their orbits. Ismaël Bullialdus disagreed, arguing that the force of gravity must be inversely proportional to the square of the distance. In a lecture to the Royal Society in 1666, Robert Hooke said that that all bodies move in a straight line unless acted on by a force, all bodies attract each other gravitationally, and the force of gravity decreases with distance by a formula that 'I own I have not discovered'. In 1679 he settled on an inverse square law for the attraction, and wrote Newton about it.[3] So Hooke was distinctly miffed when exactly the same thing appeared in *Principia*, even though Newton credited him, along with Halley and Christopher Wren.

Hooke did accept that only Newton had deduced that closed orbits are elliptical. Newton knew that the inverse square law also permits parabolic and hyperbolic orbits, but these aren't closed curves, so the motion doesn't repeat periodically. Orbits of those kinds also have astronomical applications, mainly to comets.

Newton's law goes beyond Kepler's because of one further feature, a prediction rather than a theorem. Newton realised that since the Earth attracts the Moon, it seems reasonable that the Moon should also attract the Earth. They're like two country dancers, holding hands and whirling round and round. Each dancer feels the force exerted by the other, tugging at their arms. Each dancer is held in place by that force: if they let go, they will spin off across the dance floor. However, the Earth is much more massive than the Moon, so it's like a fat man dancing with a small child. The man seems to spin in place as the child whirls round and round. But look carefully, and you'll see that the fat man is whirling too: his feet go round in a small circle, and the centre about which he rotates is slightly closer to the child than it would have been if he were spinning alone.

This reasoning led Newton to propose that *every* body in the universe attracts every other body. Kepler's laws apply to only two bodies, Sun and planet. Newton's law applies to any system of bodies whatsoever, because it provides both the magnitude and the direction of *all of the forces that occur*. Inserted into the laws of motion, the combination of all these forces determines each body's acceleration, hence velocity, hence position at any moment. The enunciation of a universal law of gravity was an epic moment in the history and development of science, revealing hidden mathematical machinery that keeps the universe ticking.

✦

Newton's laws of motion and gravity triggered a lasting alliance between astronomy and mathematics, leading to much of what we now know about the cosmos. But even when you understand what the laws are, it's not straightforward to apply them to specific problems. The gravitational force, in particular, is 'nonlinear', a technical term whose main implication is that you can't solve the equations of motion using nice formulas. Or nasty ones, for that matter.

Post-Newton, mathematicians got round this obstacle either by working with very artificial (though intriguing) problems, such as three identical masses arranged in an equilateral triangle, or by deriving approximate solutions to more realistic problems. The second approach is more practical, but actually a lot of useful ideas came from the first, artificial though it was.

For a long time, Newton's scientific heirs had to perform their calculations by hand, often a heroic task. An extreme example is Charles-Eugène Delaunay, who in 1846 started to calculate an approximate formula for the motion of the Moon. The feat took over twenty years, and he published his results in two volumes. Each has more than 900 pages, and the second volume consists entirely of the formula. In the late twentieth century his answer was checked using computer algebra (software systems that can manipulate formulas, not just numbers). Only two tiny errors were found, one a consequence of the other. Both have a negligible effect.

The laws of motion and gravity are of a special kind, called differential equations. Such equations specify the rate at which quantities change as time passes. Velocity is the rate of change of position, acceleration the rate of change of velocity. The rate at which a quantity is currently changing lets you project its value into the future. If a car is travelling at ten metres per second, then one second from now it will have moved ten metres. This type of calculation requires the rate of change to be constant, however. If the car is accelerating, then one second from now it will have moved more than ten metres. Differential equations get round this problem by specifying the instantaneous rate of change. In effect, they work with very small intervals of time, so that the rate of change can be considered constant during that time interval. It actually took mathematicians several hundred years to make sense of that idea in full logical rigour, because no finite period of time can be instantaneous unless it's zero, and nothing changes in zero time.

Computers created a methodological revolution. Instead of

calculating approximate formulas for the motion, and then putting the numbers into the formulas, you can work from the beginning with the numbers. Suppose you want to predict where some system of bodies – say the moons of Jupiter – will be in a hundred years' time. Start from the initial positions and motions of Jupiter, its moons, and any other bodies that might be important, such as the Sun and Saturn. Then, tiny time step by tiny time step, compute how the numbers describing *all* the bodies change. Repeat until you reach a hundred years: stop. A human with pencil and paper couldn't use this method on any realistic problem. It would take lifetimes. With a fast computer, however, the method becomes entirely feasible. And modern computers are very fast indeed.

It's not *quite* that easy, to be honest. Although the error at each step (caused by assuming a constant rate of change when actually it varies a little) is very small, you have to use an awful lot of steps. A big number times a small error need not be small, but carefully concocted methods keep the errors under control. The branch of mathematics known as numerical analysis is aimed at just this issue. It's convenient to refer to such methods as 'simulations', reflecting the crucial role of the computer. It's important to appreciate that you can't solve a problem merely by 'putting it on the computer'. Someone has to program the machine with the mathematical rules that make its computations match reality.

So exquisitely accurate are those rules that astronomers can predict eclipses of the Sun and Moon to the second, and predict within a few kilometres whereabouts on the planet they will occur, hundreds of years into the future. These 'predictions' can also be run *backwards* in time to pin down exactly when and where historically recorded eclipses occurred. These data have been used to date observations made thousands of years ago by Chinese astronomers, for example.

✦

Even today, mathematicians and physicists are discovering new and unexpected consequences of Newton's law of gravity. In 1993 Cris Moore used numerical methods to show that three bodies with identical masses can chase each other repeatedly along the same figure-8 shaped orbit, and in 2000 Carles Simó showed numerically that this orbit is stable, except perhaps for a slow drift. In 2001 Alain Chenciner and Richard Montgomery gave a rigorous proof that this orbit exists, based on the principle of least action, a fundamental theorem in classical mechanics.[4] Simó has discovered many similar 'choreographies', in which several bodies with the same mass pursue each other along exactly the same (complicated) path.[5]

The stability of the figure-8 three-body orbit seems to persist if the masses are slightly different, opening up a small possibility that three real stars might behave in this remarkable way. Douglas Heggie estimates that there might be one triple system of this kind per galaxy, and there's a fair chance of at least one somewhere in the universe.

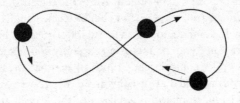

The Fig-8 three-body orbit.

These orbits all exist in a plane, but there's a novel three-dimensional possibility. In 2015 Eugene Oks realised that unusual orbits of electrons in 'Rydberg quasimolecules' might also occur in Newtonian gravity. He showed that a planet can be batted back and forth between the two stars of a binary system in a corkscrew orbit that spirals around the line that joins them.[6] The spirals are loose in the middle but tighten up near the stars. Think of joining the stars by a rotating slinky, stretched in the middle and doubling back on itself at the ends. For stars with different masses, the slinky should be tapered like a cone. Orbits like this can be stable, even if the stars don't move in circles.

Collapsing gas clouds create planar orbits, so a planet is unlikely to form in a such an orbit. But a planet or asteroid perturbed into in a highly tilted orbit might rarely be captured by the binary stars and end up corkscrewing between them. There's tentative evidence that Kepler-16b, a planet orbiting a distant star, might be one of them.

✦

One aspect of Newton's law bothered the great man himself; in fact, it bothered him more than it did most of those who built on his work. The law describes the force that one body exerts on another, but it does not address *how* the force works. It postulates a mysterious 'action at a distance'. When the Sun attracts the Earth, somehow the Earth must 'know' how far it is from the Sun. If, for example, some kind of elastic string joined the two, then the string could propagate the force, and the physics of the string would govern how strong the force was. But between Sun and Earth is only empty space. How does the Sun know how hard to pull the Earth – or the Earth know how hard to be pulled?[7]

Pragmatically, we can apply the law of gravity without worrying about a physical mechanism to transmit the force from one body to another. On the whole, that's what everyone did. A few scientists, however, possess a philosophical streak, a spectacular example being Albert Einstein. His special theory of relativity, published in 1905, changed physicists' view of space, time, and matter. Its extension in 1915 to general relativity changed their view of gravity, and almost as a side issue resolved the thorny question of how a force could act at a distance. It did so by getting rid of the force.

Einstein deduced special relativity from a single fundamental principle: the speed of light remains unchanged even when the observer is moving at a constant speed. In Newtonian mechanics, if you're in an open-top car and you throw a ball in the direction the car is moving, then the speed of the ball as measured by a stationary observer at the roadside will be the speed of the ball relative to the car, *plus* the speed of the car. Similarly, if you shine a torch beam ahead of the car, the speed of light as measured by someone at the side of the road ought to be its usual speed plus that of the car.

Experimental data and some thought experiments persuaded Einstein that light *isn't* like that. The observed speed of light is *the same* for the person shining the torch, and the one at the roadside. The logical consequences of this principle – which I've always felt should be called *non*-relativity – are startling. Nothing can travel faster than light.[8] As a body approaches the speed of light, it shrinks in the direction of motion, its mass increases, and time passes ever more slowly. At the speed of light – if that were possible – it would be infinitely thin, have infinite mass, and time on it would stop. Mass and energy are related: energy equals mass times the square of the speed of light. Finally, events that one observer considers to happen at the same time may not be simultaneous for another observer moving at a constant relative speed to the first.

In Newtonian mechanics, none of these weird things happens. Space is space and time is time, and never the twain shall meet. In special relativity, space and time are to some extent interchangeable, the extent being limited by the speed of light. Together they form a single spacetime continuum. Despite its strange predictions, special relativity has become accepted as the most accurate theory of space and time that we have. Most of its wilder effects only become apparent when objects are travelling very fast, which is why we don't notice them in everyday life.

The most obvious missing ingredient is gravity. Einstein spent years trying to incorporate the force of gravity into relativity, motivated in part by an anomaly in the orbit of Mercury.[9] The end result was general relativity, which extends the formulation of special relativity from a 'flat' spacetime continuum to a 'curved' one. We can get a rough understanding of what's involved by cutting space down to two dimensions instead of three. Now space becomes a plane, and special relativity describes the motion of particles in this plane. In the absence of gravity, they follow straight lines. As Euclid pointed out, a straight line is the shortest distance between two points. To put gravity into the picture, place a star in the plane. Particles no longer follow straight lines; instead, they orbit the star along curves, such as ellipses.

In Newtonian physics, these paths are curved because a force diverts the particle from a straight line. In general relativity, a similar effect is obtained by bending spacetime. Suppose the star distorts the plane, creating a circular valley – a 'gravity well' with the star at the bottom – and assume that moving particles follow whichever path is shortest. The technical term is *geodesic*. Since the spacetime continuum is bent, geodesics are no longer straight lines. For example, a particle can be trapped in the valley, going round and round at a fixed height, like a planet in a closed orbit.

Instead of a hypothetical force that causes the particle's path to curve, Einstein substituted a spacetime that's *already* curved, and whose curvature affects the path of a moving particle. No action at

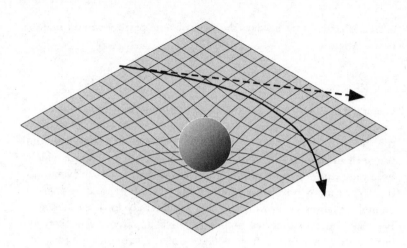

Effect of curvature/gravity on a particle passing a star or planet.

a distance is needed: spacetime is curved because that's what stars do to it, and orbiting bodies respond to nearby curvature. What we and Newton refer to as gravity, and think of as a force, is actually the curvature of spacetime.

Einstein wrote down mathematical formulas, the Einstein field equations,[10] which describe how curvature affects the motion of masses, and how the mass distribution affects curvature. In the absence of any masses, the formula reduces to special relativity. So all of the weird effects, such as time slowing down, also happen in general relativity. Indeed, gravity can *cause* time to slow down, even for an object that is not moving. Usually these paradoxical effects are small, but in extreme circumstances the behaviour that relativity (of either type) predicts differs significantly from Newtonian physics.

Think this all sounds mad? Many did, to begin with. But anyone who uses satnav in their car is relying on both special and general relativity. The calculations that tell you you're on the outskirts of Bristol heading south on the M32 motorway rely on timing signals from orbiting satellites. The chip in your car that computes your location has to correct those timings for two effects: the speed with which the satellite is moving, and its position in Earth's gravity well. The first requires special relativity, the second general relativity. Without these corrections, within a few days the satnav would place you in the middle of the Atlantic.

✦

General relativity shows that Newtonian physics is *not* the true, exact 'system of the world' that he (and almost all other scientists prior to the twentieth century) believed it to be. However, that discovery did not spell the end of Newtonian physics. In fact, it's far more widely used now, and for more practical purposes, than it was in Newton's day. Newtonian physics is simpler than relativity, and it's 'good enough for government work', as they say: literally so. The differences between the two theories mainly become apparent when considering exotic phenomena such as black holes. Astronomers and space mission engineers, mainly employed by governments or contracts between governments and organisations such as NASA and ESA, still use Newtonian mechanics for almost all calculations. There are a few exceptions where timing is delicate. As the story unfolds, we'll see the influence of Newton's law of gravity over and over again. It really is that important: one of the greatest scientific discoveries of all time.

However, when it comes to cosmology – the study of the entire universe and especially its origins – we must ditch Newtonian physics. It can't explain the key observations. Instead, general relativity must be invoked, ably assisted by quantum mechanics. And even those two great theories seem to need extra help.

Collapse of the Solar Nebula

> Two thousand million years or so ago two galaxies were colliding,
> or rather, were passing through each other... At about the same
> time – within the same plus-or-minus ten percent margin of error,
> it is believed – practically all of the suns of both of those galaxies
> became possessed of planets.
>
> Edward E. Smith PhD, *Triplanetary*

TRIPLANETARY IS THE FIRST in Edward E. Smith's celebrated 'Lensman'
series of science fiction novels, and its opening paragraph reflects a
theory about the origin of planetary systems that was in vogue when
the book appeared in 1948. Even today it would be a powerful way to
start a science fiction novel; back then, it was breathtaking. The novels
themselves are early examples of 'widescreen baroque' space opera,
a cosmic battle between the forces of good (represented by Arisia)
and evil (Eddore) that takes six books to complete. The characters are
cardboard, the plots trite, but the action is enthralling, and at the time
the scope was unparalleled.

Today we no longer think that galactic collisions are needed to
create planets, though astronomers do see them as one of the four main
ways to make stars. The current theory of the formation of our own
solar system, and many other planetary systems, is different, yet no less
breathtaking than that opening paragraph. It goes like this.

Four and a half billion[1] years ago, a cloud of hydrogen gas six
hundred trillion kilometres across began to tear itself slowly to pieces.
Each piece condensed to create a star. One such piece, the solar nebula,
formed the Sun, together with its solar system of eight planets, five (so

far) dwarf planets, and thousands of asteroids and comets. The third rock from the Sun is our homeworld: Earth.

Unlike the fiction, it might even be true. Let's examine the evidence.

✦

The idea that the Sun and planets all condensed from a vast cloud of gas appeared remarkably early, and for a long time it was the prevailing scientific theory of their origins. When problems emerged, it went out of favour for nearly 250 years, but it has now been revived, thanks to new ideas and new data.

René Descartes is mainly famous for his philosophy – 'I think, therefore I am' – and his mathematics, notably coordinate geometry, which translates geometry into algebra and vice versa. But in his day 'philosophy' referred to many areas of intellectual activity, including science, which was *natural* philosophy. In his 1664[2] *Le Monde* (The World) Descartes tackled the origin of the solar system. He argued that initially the universe was a formless jumble of particles, circulating like whirlpools in water. One unusually large vortex swirled ever more tightly, contracting to form the Sun, and smaller vortices around it made the planets.

At a stroke, this theory explained two basic facts: why the solar system contains many separate bodies, and why the planets all go round the Sun in the same direction. Descartes's vortex theory doesn't agree with what we now know about gravity, but Newton's law wouldn't appear for another two decades. Emanuel Swedenborg replaced Descartes's swirling vortices with a huge cloud of gas and dust in 1734. In 1755 the philosopher Immanuel Kant gave the idea his blessing; the mathematician Pierre-Simon de Laplace stated it independently in 1796.

All theories of the origin of the solar system must explain two key observations. An obvious one is that matter has clumped together into discrete bodies: Sun, planets, and so on. A subtler one concerns a quantity known as *angular momentum*. This emerged from mathematical investigations into the deep implications of Newton's laws of motion.

The related concept of momentum is easier to understand. It governs the tendency of a body to travel at a fixed speed in a straight line when no forces are acting, as Newton's first law of motion states. Sports commentators use the term metaphorically: 'She's got the momentum now.' Statistical analysis offers remarkably little support for the proposal that a series of good scores tends to lead to more; commentators explain away the failures of their metaphor by observing (after the event) that the momentum has been lost again. In mechanics, the mathematics of moving bodies and systems, momentum has a very specific meaning, and one consequence is that you *can't* lose it. All you can do it transfer it to something else.

Think of a moving ball. Its speed tells us how fast it's moving: 80 kilometres per hour, say. Mechanics focuses on a more important quantity, velocity, which measures not just how fast it's going, but in which direction. If a perfectly elastic ball bounces off a wall, its speed remains unchanged but its velocity reverses direction. Its momentum is its mass multiplied by its velocity, so momentum also has both a size and a direction. If a light body and a massive one both move at the same speed in the same direction, the massive one has more momentum. Physically, it's then necessary to apply more force to change how the body is moving. You can easily bat away a ping-pong ball passing at 50 kph, but no one in their right mind would try that with a truck.

Mathematicians and physicists like momentum because, unlike velocity, it's conserved as a system changes over time. That is, the total momentum of the system remains fixed at whatever size, and direction, it was to start with.

That may sound implausible. If a ball hits a wall and bounces, its momentum changes direction, so it's not conserved. But the wall, much more massive, bounces a tiny bit too, and it bounces *the other way*. After that, other factors come into play, such as the rest of the wall, and I've kept up my sleeve the get-out clause: the conservation law only works when there are no external forces, that is, outside interference. This is how a body can acquire momentum to begin with: something gives it a shove.

Angular momentum is similar, but it applies to bodies that are rotating rather than moving in a straight line. Even for a single particle, its definition is tricky, but like momentum it depends both on the mass of the particle and its velocity. The main new feature is that it also depends on the axis of rotation – the line about which the particle is considered to be rotating. Imagine a spinning top. It spins around the line that runs through the middle of the top, so every particle of matter in the top rotates around this axis. The particle's angular momentum about that axis is its rate of spin multiplied by its mass. But the direction in which the angular momentum points is *along the spin axis*. That is, at right angles to the plane in which the particle is rotating. The angular momentum of the entire top, again considered about its axis, is obtained by adding together all of the angular momenta of its constituent particles, taking direction into account when necessary.

The size of a spinning system's total angular momentum tells us how strongly it's spinning, and the direction of the angular momentum tells us which axis it's spinning about, on average.[3] Angular momentum is conserved in any system of bodies subject to no external twisting forces (jargon: torque).

✦

This useful little fact has immediate implications for the collapse of a gas cloud: some good, some bad.

The good one is that, after some initial confusion, the gas molecules tend to spin in a single plane. Initially, each molecule has a certain amount of angular momentum about the centre of gravity of the cloud. Unlike a top, a gas cloud is not rigid, so these speeds and directions probably vary wildly. It's unlikely that all these quantities will cancel out perfectly, so initially the total angular momentum of the cloud is non-zero. The total angular momentum therefore points in some definite direction, and has a definite size. Conservation tells us that as the gas cloud evolves under gravity, its total angular momentum *doesn't change*. So the direction of the axis stays fixed, frozen in at the moment the cloud first formed. And the size – the total amount of spin, so to speak – is also frozen in. What can change is the distribution of the gas molecules. Every molecule of gas exerts a gravitational attraction on every other molecule, and the initially chaotic globular gas cloud collapses to form a flat disc, spinning about the axis like a plate on a pole in a circus.

This is good news for the solar nebula theory, because all the planets of the solar system have orbits that lie very close to the same plane – the ecliptic – and they all revolve in the same direction. That's why early astronomers guessed that the Sun and planets all condensed from a cloud of gas, after it had collapsed to create a protoplanetary disc.

Unfortunately for this 'nebular hypothesis', there is also some bad news: 99% of the solar system's angular momentum resides in the planets, with only 1% in the Sun. Although the Sun contains virtually the entire mass of the solar system, it's spinning quite slowly and its particles are relatively close to the central axis. The planets, though lighter, are much further away and move much faster, so they hog nearly all of the angular momentum.

However, detailed theoretical calculations show that a collapsing gas cloud doesn't do that. The Sun gobbles up most of the matter in the entire gas cloud, including a lot that was originally much further from the centre. So you'd expect it to have gobbled up the lion's share of the angular momentum ... which it spectacularly failed to do. Nevertheless, the current allocation of angular momentum, in which the planets get the lion's share, is entirely consistent with the dynamics of the solar system. It *works*, and it has done so for billions of years. There's no logical problem with the dynamics as such: just with how it all got started.

✦

One potential way out of this dilemma quickly emerged. Suppose the Sun formed *first*. Then it did gobble up pretty much all of the angular momentum in the gas cloud, because it gobbled up pretty much all of the gas. Afterwards it could acquire planets by *capturing* lumps of matter that passed nearby. If they were far enough away from the Sun, and moving at the right speed to be captured, the 99% figure would work out, just as it does today.

The main problem with this scenario is that it's very tricky to capture a planet. Any would-be planet that comes close enough will speed up as it approaches the Sun. If it manages not to fall into the Sun, it will swing round the Sun and be flung back out again. Since it's hard to capture just one planet, what price eight?

Perhaps, Count Buffon mused in 1749, a comet crashed into the Sun and splashed off enough material to create the planets. No, said Laplace in 1796: planets formed like that would eventually fall back into the Sun. The reasoning is similar to the 'no capture' line of argument, but in reverse. Capture is tricky because what comes down must go up again (unless it hits the Sun and gets engulfed). Splashing off is tricky because what goes up must come down. In any case we now know (as they didn't then) that comets are too lightweight to make a planet-sized splash, and the Sun is made of the wrong stuff.

In 1917 James Jeans suggested the tidal theory: a wandering star passed near the Sun and sucked out some of its material in a long, thin cigar. Then the cigar, which was unstable, broke into clumps that became the planets. Again, the Sun has the wrong composition; moreover this proposal requires a remarkably unlikely near collision, and doesn't endow the outermost planets with enough angular momentum to stop them falling back. Dozens of theories – all different, but all variations on similar themes – were proposed. Each fits some facts, and struggles to explain others.

By 1978, the apparently discredited nebular model was back in vogue. Andrew Prentice came up with a plausible solution to the angular momentum problem – recall: the Sun has way too little, the planets have too much. What's needed is a way to prevent angular momentum being conserved: to gain some or lose some. Prentice suggested that grains of dust concentrate near the centre of the disc of gas, and friction with those grains slows the rotation of the newly condensed Sun. Victor Safronov developed similar ideas around the same time, and his book on the topic led to the widespread adoption of the 'collapsing disc' model, in which the Sun and planets (and much else) all condensed out of a single massive gas cloud, pulled apart into clumps of many different sizes by its own gravity, modified by friction.

This theory has the merit of explaining why the inner planets (Mercury, Venus, Earth, Mars) are mainly rocky, while the outer ones (Jupiter, Saturn, Uranus, Neptune) are gas and ice giants. The lighter elements in the protoplanetary disc would accumulate further out than the heavy ones, though with much turbulent mixing. The prevailing theory for the giants is that a rocky core formed first, and its gravity attracted hydrogen, helium, and some water vapour, plus relatively small amounts of other material. However, models of planet formation have struggled to reproduce this behaviour.

In 2015 Harold Levison, Katherine Kretke, and Martin Duncan performed computer simulations supporting an alternative option: the cores slowly accreted from 'pebbles', lumps of rocky matter up to a metre across.[4] In theory this process can build up a core ten times the mass of the Earth in a few thousand years. Previous simulations had thrown up a different problem with this idea: it generates hundreds of planets the size of the Earth. The team showed that this problem is avoided if the pebbles come into being slowly enough to interact with each other gravitationally. Then the bigger ones scatter the rest out of the disc. Simulations with different parameters often produced between one and four gas giants at a distance of 5–15 AU from the Sun, consistent with the present structure of the solar system. One astronomical unit (AU) is the distance from the Earth to the Sun, and this is often a convenient way to grasp relatively small cosmic distances.

A good way to test the nebular model is to find out whether similar processes are going on elsewhere in the cosmos. In 2014 astronomers captured a remarkable image of the young star HL Tauri, 450 light years away in the constellation Taurus. The star is surrounded by concentric bright rings of gas, with dark rings in between. The dark rings are almost certainly caused by nascent planets sweeping up dust and gas. It would be difficult to find a more dramatic confirmation of a theory.

Atacama Large Millimeter Array image of HL Tauri, showing concentric rings of dust and gaps between them.

✦

It's easy to believe that gravity can cause things to clump, but how can it also pull them apart? Let's develop some intuition. Again, some serious mathematics, which we won't do here, confirms the general gist. I'll start with clumping.

A body of gas whose molecules attract each other *gravitationally* is very different from our usual experience of gases. If you fill a room with gas, it will very rapidly disperse so that the gas has the same density everywhere. You don't find odd pockets in your dining room where there's no air. The reason is that the molecules of air bounce around at random, and would quickly occupy any spare space. This behaviour is enshrined in the famous second law of thermodynamics, whose usual interpretation is that a gas is as disordered as possible. 'Disordered', in this context, has the connotation that everything should be thoroughly mixed together; that means that no region should be denser than any other.

To my mind this concept, technically known as entropy, is far too slippery to be captured by a simple word like 'disorder' – if only because 'evenly mixed' sounds to me like an *ordered* state. But for the moment, I'm going to toe the orthodox line. The mathematical formulation doesn't actually mention order or disorder at all, but it's too technical to discuss right now.

What holds in a room surely holds in a large room, so why not in a room the size of the entire universe? Indeed, in the universe itself? Surely the second law of thermodynamics implies that all of the gas in the universe ought to spread itself out uniformly into some sort of thin fog?

If that were true, it would be bad news for the human race, because we're not made from thin fog. We're distinctly lumpy, and we live on a rather bigger lump that orbits a lump so big that it sustains energetic nuclear reactions, producing heat and light. Indeed, people who dislike the usual scientific descriptions of the origin of humanity often invoke the second law of thermodynamics to 'prove' that we wouldn't be able to exist unless some hyper-intelligent being had deliberately manufactured us and arranged the universe to suit our requirements.

However, the thermodynamic model of gas in a room is not appropriate for working out how the solar nebula, or the entire universe, should behave. It has the wrong kind of interactions. Thermodynamics assumes that molecules notice each other only when they collide; then they bounce off each other. The bounces are perfectly elastic, meaning that no energy is lost, so the molecules continue bouncing forever. Technically, the forces that govern the interactions of molecules in a thermodynamic model of a gas are short range and repulsive.

Imagine a party where everyone is blindfolded and has their ears plugged, so that the only way to discover anyone else is present is to bump into them. Imagine, moreover, that everyone is hugely antisocial, so that when they do encounter someone else, both people immediately push each other away. It's plausible that after some initial bumping and wobbling they spread themselves around fairly evenly. Not all the time, because sometimes they come close together by accident, or even collide, but on average they stay spread out. A thermodynamic gas is like that, with absolutely gigantic numbers of molecules acting as people.

A gas cloud in space is more complicated. The molecules still bounce if they hit, but there's a second type of force: gravity. Gravity is ignored in thermodynamics because in that context its effects are negligibly small. But in cosmology, gravity is the dominant player, because there's an awful lot of gas. Thermodynamics keeps it gaseous, but gravity determines what the gas does on larger scales. Gravity is long range and attractive, almost the exact opposite of elastic bouncing. 'Long range' means that bodies interact even when they're far apart. The gravity of the Moon (and to a lesser extent the Sun) raises tides in the Earth's oceans, and the Moon is 400,000 kilometres away. 'Attractive' is straightforward: it causes the interacting bodies to move towards each other.

This is like a party where everyone can see everyone else from across the room – though less clearly from a distance – and as soon as they see anybody, they rush towards them. It's hardly surprising that a mass of gas interacting under gravity naturally becomes clumpy. In very small regions of the clumps, the thermodynamic model dominates, but on a larger scale the tendency to cluster dominates the dynamics.

If we're trying to work out what would happen to a hypothetical solar nebula, on the scale of solar systems or planets, we have to think about the long-range attractive force of gravity. The short-range repulsion between molecules that collide might tell us something about the state of a small region in a planet's atmosphere, but it won't tell us about the *planet*. In fact, it will mislead us into imagining that the planet should never have formed.[5]

Clumpiness is an inevitable consequence of gravity. A uniform spread is not.

✦

Since gravity causes matter to clump together, how can it also rip a molecular cloud apart? It seems contradictory.

The answer is that competing clumps can form at the same time. The mathematical arguments that support the collapse of a gas cloud into a flat, spinning disc assume we start with a region of gas that is roughly spherical – maybe shaped like an American football, but not like a dumbbell. However, a large region of gas will have occasional, randomly located places where by chance the matter is a little denser than elsewhere. Each such region acts as a centre, attracting more matter from its surroundings and exerting an ever-stronger gravitational force. The resulting cluster starts out fairly spherical and then collapses to a spinning disc.

However, in a large enough region of gas, several such centres can form. Although gravity is long range, its force drops off as the distance between bodies increases. So molecules are attracted to the nearest centre. Each centre is surrounded by a region in which its gravitational pull dominates. If there are *two* very popular people at the party, in opposite corners of the room, the party will split into two groups. So the gas cloud organises itself into a three-dimensional patchwork of attracting centres. These regions tear the cloud apart along their common boundaries. In practice what happens is a bit more complicated, and fast-moving molecules can escape the influence of the nearest centre and end up near a different one, but broadly speaking this is what we expect. Each centre condenses to form a star, and some of the debris surrounding it may form planets and other smaller bodies.

This is why an initially uniform cloud of gas condenses into a whole series of separate, relatively isolated star systems. Each system corresponds to one of the dense centres. But even then it's not totally straightforward. If two stars are close enough together, or approach each other for chance reasons, they can end up orbiting their common centre of mass. Now they form a binary star. In fact, systems of three or more stars can arise, loosely bound together by their mutual gravitation.

These multiple star systems, especially binary ones, are very common in the universe. The nearest star to the Sun, Proxima Centauri, is quite close (in astronomical terms) to a binary star called Alpha Centauri, whose individual stars are Alpha Centauri A and B. It seems likely that Proxima orbits both of these, but it probably takes half a million years to go once round its orbit. The distance between A and B is comparable to the distance from Jupiter to the Sun; it varies between about 11 and 36 AU.

In contrast, the distance from Proxima to either A or B is more like 15,000 AU, roughly a thousand times greater. Therefore, by inverse square law gravity, the force that A and B exert on Proxima is about one millionth of the force they exert on each other. Whether that's strong enough to keep Proxima in a stable orbit depends sensitively on what else is close enough to prise it from A and B's tenuous grasp. At any rate, we won't be around to see what happens.

✦

The early history of the solar system must have included periods of violent activity. The evidence is the huge number of craters on most bodies, especially the Moon, Mercury, Mars, and various satellites, showing that they were bombarded by innumerable smaller bodies. The relative ages of the resulting craters can be estimated statistically, because younger craters partially destroy older ones when they overlap, and most of the observed craters on these worlds are very ancient indeed. Even so, occasional new ones form, but most of these are very small.

The big problem here is to sort out the sequence of events that shaped the solar system. In the 1980s the invention of powerful computers and efficient and accurate computational methods permitted detailed mathematical modelling of collapsing clouds. Some sophistication is required because crude numerical methods fail to respect physical constraints such as conservation of energy. If this mathematical artefact causes energy to decrease slowly, the effect is like friction. Instead of following a closed orbit, a planet will spiral slowly into the Sun. Other quantities such as angular momentum must also be conserved. Methods that avoid this danger are of recent vintage. The most accurate ones are known as symplectic integrators, after a technical way to reformulate the equations of mechanics, and they conserve all relevant physical quantities *exactly*. Careful, precise simulations reveal plausible and very dramatic mechanisms for the formation of planets, which fit observations well. According to these ideas, the early solar system was very different from the sedate one we see today.

Astronomers used to think that once the solar system came into being, it was very stable. The planets trundled ponderously along preordained orbits and nothing much changed; the elderly system we see now is pretty much what it was in its youth. No longer! It's now thought that the larger worlds, the gas giants Jupiter and Saturn and the ice giants Uranus and Neptune, first appeared outside the 'frost line' where water freezes, but subsequently reorganised each other in a lengthy gravitational tug-of-war. This affected all of the other bodies, often in dramatic ways.

Mathematical models, plus a variety of other evidence from nuclear physics, astrophysics, chemistry, and many other branches of science, have led to the current picture: the planets didn't form as single clumps, but by a chaotic process of accretion. For the first 100,000 years, slowly growing 'planetesimals' swept up gas and dust, and created circular rings in the nebula by clearing out gaps between them. Each gap was littered with millions of these tiny bodies. At that point the planetesimals ran out of new matter to sweep up, but there were so many of them that they kept bumping into each other. Some broke up, but others merged; the mergers won and planets built up, piece by tiny piece.

In this early solar system, the giants were closer together than they are today, and millions of tiny planetesimals roamed the outer regions. Today the order of the giants, outwards from the Sun, is Jupiter, Saturn, Uranus, Neptune. But in one likely scenario it was originally Jupiter, Neptune, Uranus, Saturn. When the solar system was about 600 million years old, this cosy arrangement came to an end. All of the planets' orbital periods were slowly changing, and Jupiter and Saturn wandered into a 2:1 resonance – Jupiter's 'year' became exactly half that of Saturn. In general, resonances occur when two orbital or rotational periods are related by a simple fraction, here one half.[6] Resonances have a strong effect on celestial dynamics, because bodies in resonant orbits repeatedly align in exactly the same way, and I'll be saying a lot more about them later. This prevents disturbances 'averaging out' over long periods of time. This particular resonance pushed Neptune and Uranus outwards, and Neptune overtook Uranus.

This rearrangement of the larger bodies of the solar system disturbed the planetesimals, making them fall towards the Sun. All hell broke loose as planetesimals played celestial pinball among the planets. The giant planets moved out, and the planetesimals moved in. Eventually the planetesimals took on Jupiter, whose huge mass was decisive. Some planetesimals were flung out of the solar system altogether, while the rest went into long, thin orbits going out to huge distances. After that, everything mostly settled down, but the Moon, Mercury, and Mars still have battle scars resulting from the chaos.[7] And bodies of all shapes, sizes, and compositions were scattered far and wide.

Mostly settled down. It hasn't stopped. In 2008 Konstantin Batygin and Gregory Laughlin simulated the future of the solar system for 20 billion years, and the initial results revealed no serious instabilities.[8] Refining the numerical method to seek out potential instabilities, changing the orbit of at least one planet in a major way, they discovered a possible future in which Mercury hits the Sun about 1·26 billion years from now, and another in which Mercury's erratic movements eject Mars from the solar system 822 million years from now, followed by a collision between Mercury and Venus 40 million years later. The Earth sails serenely on, unaffected by the drama.

Early simulations mainly used averaged equations, not suitable for collisions, and ignored relativistic effects. In 2009 Jacques Laskar and Mickael Gastineau simulated the next 5 billion years of the solar system, using a method that avoided these problems,[9] but the results were much the same. Because tiny differences in initial conditions can have a large effect on long-term dynamics, they simulated 2,500 orbits, all starting within observational error of current conditions. In about 25 cases, the near resonance pumps up Mercury's eccentricity, leading either to a collision with the Sun, a collision with Venus, or a close encounter that radically changes the orbits of both Venus and Mercury. In one case, Mercury's orbit subsequently becomes less eccentric, causing it to destabilise all four of the inner planets within the next 3·3 billion years. The Earth is then likely to collide with Mercury, Venus, or Mars. And again there's a slight chance that Mars will be ejected from the solar system altogether.[10]

Inconstant Moon

It is the very error of the moon: She comes more nearer earth than she was wont, and makes men mad.

William Shakespeare, *Othello*

OUR MOON IS UNUSUALLY large.

It has a diameter just over one quarter of the Earth's, much larger than most other moons: so large, in fact, that the Earth–Moon system is sometimes referred to as a double planet. (Some jargon: Earth is the primary, the Moon is a satellite. Going up a level, the Sun is the primary of the planets in the solar system.) Mercury and Venus have no moons, while Mars, the planet most closely resembling Earth, has two tiny moons. Jupiter, the largest planet in the solar system, has 67 known moons, but 51 of them are less than 10 kilometres across. Even the largest, Ganymede, is less than one-thirtieth the size of Jupiter. Saturn is the most prolific in the satellite stakes, with over 150 moons and moonlets and a giant and complex ring system. But its largest moon, Titan, is only one twentieth as large as its primary. Uranus has 27 known moons, the largest being Titania, less than 1600 kilometres across. Neptune's only large moon is Triton, about one-twentieth of the planet's size; in addition astronomers have found 13 very small moons. Among the worlds of the solar system, only Pluto does better than us: four of its satellites are tiny, but the fifth, Charon, is about half the size of its primary.

The Earth–Moon system is unusual in another respect: its unusually large angular momentum. Dynamically, it has more 'spin' than it ought to. There are other surprises about the Moon, too, and we'll come to those in due course. The Moon's exceptional nature adds weight to a natural question: how did the Earth acquire its satellite?

The theory that fits the current evidence best is dramatic: the giant impact hypothesis. Early in its formation, our home planet was about 10% smaller than it is now, until a body about the size of Mars smashed into it, splashing off huge amounts of matter – initially much of it molten rock, in globules of all sizes, many of which merged as the rock began to cool. Part of the impactor united with the Earth, which became larger. Part of it became the Moon. The rest was dispersed elsewhere in the solar system.

Mathematical simulations support the giant impact scenario, while other theories fare less well. But in recent years, the giant impact hypothesis has started to run into trouble, at least in its original version. The origin of the Moon may still be up for grabs.

✦

The simplest theory is that the Moon accreted from the solar nebula, along with everything else, during the formation of the solar system. There was a lot of debris, in a huge range of sizes. As it started to settle down, larger masses grew by attracting smaller ones that merged with them after collisions. The planets formed this way, asteroids formed this way, comets formed this way, and moons formed this way. So presumably our Moon formed this way.

If so, however, it didn't form anywhere near its present orbit. The killer is angular momentum: there's too much of it. Another problem is the Moon's composition. As the solar nebula condensed, different elements were abundant at different distances. The heavier stuff stayed near the Sun, while radiation blew the lighter elements further out. That's why the inner planets are rocky, with iron–nickel cores, but the outer ones are mainly gas and ice – which is gas that got so cold, it froze. If Earth and Moon formed at roughly the same distance from the Sun, and at roughly the same time, they ought to have similar rocks in similar proportions. But the Moon's iron core is far smaller than the Earth's. In fact, Earth's total proportion of iron is eight times as large as the Moon's.

In the 1800s Charles Darwin's son George came up with another theory: in its early days the Earth, still molten, was spinning so fast that part of it broke off under the action of centrifugal force. He did the sums using Newtonian mechanics, and predicted that the Moon must be moving away from the Earth, which turns out to be true. This event would have left a large scar, and there was an obvious candidate: the Pacific Ocean. However, we now know that Moon rock is much older than the oceanic crustal material in the Pacific. That rules out the Pacific basin, but not necessarily Darwin's fission theory.

Plenty of other scenarios have been suggested, some rather wild. Perhaps a natural nuclear reactor (at least one is known to have existed, by the way[1]) went critical, exploded, and expelled the lunar material. If the reactor was near the boundary between mantle and core, close to the equator, a lot of Earth rock would have gone into equatorial orbit. Or perhaps Earth originally had two moons, which collided. Or we stole a moon from Venus, which neatly explains why Venus doesn't have one. Though it fails to explain why, if that theory is correct, Earth originally didn't.

A less dramatic alternative is that the Earth and the Moon formed separately, but later the Moon came close enough to the Earth to be captured by its gravity. This idea has several things going for it. The Moon has the right size, and it's in a sensible orbit. Moreover, capture explains why the Moon and Earth are 'tidally locked' by their mutual gravity, so that the same side of the Moon always faces the Earth. It wobbles a bit (jargon: libration), but that's normal with tidal locking.

The main issue is that although gravitational capture sounds reasonable (bodies attract each other, after all) it's actually rather unusual. The motion of celestial bodies involves hardly any friction – there's some, for example with the solar wind, but its dynamic effects are minor – so energy is conserved. The (kinetic) energy that a 'falling' body acquires as it approaches another one, pulled by their mutual gravitational interaction, is therefore big enough for the body to *escape* that pull again. Typically the two bodies approach, swing round each other, and separate.

Alternatively, they collide.

Clearly the Earth and Moon did neither.

There are ways round this problem. Perhaps the early Earth had a huge extended atmosphere, which slowed the Moon down when it came close, without breaking it up. There's a precedent: Neptune's moon Triton is exceptional not just in its size, compared with the planet's other moons, but in its direction of motion, which is 'retrograde' – the opposite way round to most of the bodies of the solar system, including all of the planets. Astronomers think that Triton was captured by Neptune. Originally, Triton was a Kuiper belt object (KBO), the name given to a swarm of smallish bodies orbiting beyond Neptune. This is an origin that it probably shares with Pluto. If so, captures do occur.

Another observation constrains the possibilities even more. Although the overall geological compositions of the Earth and Moon are very different, the detailed composition of the Moon's surface rocks is remarkably similar to that of the Earth's mantle. (The mantle lies between the continental crust and the iron core.) Elements have 'isotopes' that are chemically almost identical, but differ in the particles forming the atomic nucleus. The commonest oxygen isotope, oxygen-16, has eight protons and eight neutrons. Oxygen-17 has an extra neutron, and oxygen-18 has a second extra neutron. When rocks form, oxygen is incorporated through chemical reactions. Samples of Moon rock brought back by *Apollo* astronauts have the same ratios of oxygen and other isotopes as the mantle.

In 2012 Randall Paniello and coworkers analysed zinc isotopes in lunar material, finding that it has less zinc than the Earth, but a higher proportion of heavy isotopes of zinc. They concluded that the Moon has lost zinc by evaporation.[2] Again, in 2013 a team under Alberto Saal reported that atoms of hydrogen included in lunar volcanic glass and olivine have very similar isotope ratios to Earth's water. If the Earth and Moon originally formed separately, it would be unlikely for their isotope ratios to be so alike.

The simplest explanation is that these two bodies have a common origin, despite differences in their cores. However, there's an alternative: perhaps they had distinct origins, and their composition was different when they formed, but later they were mixed together.

✦

Let's review the evidence that needs explaining. The Earth–Moon system has unusually large angular momentum. The Earth has a lot less iron than the Moon, yet the lunar surface has very similar isotope ratios to the Earth's mantle. The Moon is unusually large, and tidally locked to its primary. Any viable theory has to explain, or at least be consistent with, these observations to be remotely plausible. And *none* of the simple theories does that. It's the Sherlock Holmes cliché: 'When you have eliminated the impossible, then whatever remains, however unlikely, must be the truth.' And the simplest explanation that does fit the evidence is something that, until the late twentieth century, astronomers would have rejected because it seemed so improbable. Namely, Earth collided with something else, so massive that the collision melted both bodies. Some of the molten rock splashed off to form the Moon, and what merged with the Earth contributed much of its mantle.

This giant impact hypothesis, in its currently favoured incarnation, dates from 1984. The impactor even has a name: Theia. However, unicorns have a name but don't exist. If Theia ever existed, the only remaining traces are on the Moon and deep in the Earth, so the evidence has to be indirect.

Few ideas are truly original, and this one goes back at least to Reginald Daly, who objected to Darwin's fission theory because when you do the sums properly, the Moon's current orbit doesn't trace all the way back to the Earth when you run time backwards. An impact, Daly proposed, would work a lot better. The main apparent problem, at that time, was: impact with what? In those days, astronomers and mathematicians thought that the planets formed in pretty much their present orbits. But as computers got more powerful, and the implications of Newton's mathematics could be explored in more realistic settings, it became clear that the early solar system kept changing dramatically. In 1975 William Hartmann and Donald Davis performed calculations suggesting that after the planets had formed, several smaller bodies were left over. These might be captured and become moons, or they might collide, either with each other or with a planet. Such a collision, they said, could have created the Moon, and is consistent with many of its known properties.

In 1976 Alastair Cameron and William Ward proposed that another planet, about the size of Mars, collided with the Earth, and some of the material that splashed off aggregated to form the Moon.[3] Different constituents would behave differently under the massive forces and heat generated by the impact. Silicate rock (on either body) would vaporise, but the Earth's iron core, and any metallic core that the impactor might possess, would not. So the Moon would end up with much less iron than the Earth, but the surface rocks of the Moon and the Earth's mantle, condensing back from the vaporised silicates, would be extremely similar in composition.

In the 1980s Cameron and various coworkers carried out computer simulations of the consequences of such an impact, showing that a Mars-sized impactor – Theia – fits observations best.[4] At first it seemed plausible that Theia could splash off a chunk of the Earth's mantle, while contributing very little of its own material to the rocks that became the Moon. That would explain the very similar composition of these two types of rock. Indeed, this was seen as strong confirmation of the giant impact hypothesis.

Until a few years ago, most astronomers accepted this idea. Theia smashed into the primeval Earth very soon (in cosmological terms) after the formation of the solar system, between 4·5 and 4·45 billion years ago. The two worlds didn't collide head-on, but at an angle of about 45 degrees. The collision was relatively slow (again in cosmo-logical terms): around 4 kilometres per second. Calculations show that if Theia had an iron core, this would have merged with the main body of the Earth, and being denser than the mantle, it would have sunk and coalesced with the Earth's core; remember, the rocks were all molten at this stage. That explains why Earth has a lot more iron than the Moon. About one fifth of Theia's mantle, and quite a lot of Earth's silicate rock, was thrown into space. Half of that ended up orbiting the Earth, and aggregated to create the Moon. The other half escaped from the Earth's gravity and orbited the Sun. Most of it stayed in orbits roughly similar to the Earth's, so it collided with the Earth or the newly formed Moon. Many of the lunar craters were created by these secondary impacts. On Earth, however, erosion and other processes erased most impact craters.

The impact gave the Earth extra mass, and a lot of extra angular momentum: so much that it spun once every five hours. The Earth's slightly oblate shape, squashed at the poles, exerted tidal forces that aligned the Moon's orbit with the Earth's equator, and stabilised it there.

Measurements show that the crust of the Moon on the side that is now turned away from Earth is thicker. The thinking is that some of the splashed-off material in Earth orbit initially failed to be absorbed into what became the Moon. Instead, a second, smaller moon collected together in a so-called 'Lagrange point', in the same orbit as the Moon but 60 degrees further along it (see Chapter 5). After 10 million years, as both bodies drifted slowly away from the Earth, this location became unstable and the smaller moon collided with the larger one. Its material spread across the far side of the Moon, thickening the crust.

✦

I've used the words 'simulation' and 'calculation' quite a lot, but you can't do a sum unless you know what you want to calculate, and you can't simulate something by just 'putting it on the computer'. Someone has to set up the calculation, in exquisite detail; someone has to write software that tells the computer how to do the sums. These tasks are seldom straightforward.

Simulating a cosmic impact is a horrendous computational problem. The matter involved can be solid, liquid, or vapour, and different physical rules apply to each case, requiring different mathematical formulations. At least four types of matter are involved: core and mantle for each of Theia and Earth. The rocks, in whatever state, can fragment or collide. Their motion is governed by 'free boundary conditions', meaning that the fluid dynamics does not take place in a fixed region of space with fixed walls. Instead, the fluid itself 'decides' where its boundary is, and its location changes as the fluid moves. Free boundaries are much harder to handle than fixed ones, both theoretically and computationally. Finally, the forces that act are gravitational, so they are nonlinear. That is, instead of changing proportionately to distance, they change according to the inverse square law. Nonlinear equations are notoriously harder than linear ones.

Traditional pencil-and-paper mathematical methods can't hope to solve even simplified versions of the problem. Instead, fast computers with lots of memory use numerical methods to approximate the problem, and then do a lot of brute force calculations to get an approximate answer. Most simulations model the colliding bodies as droplets of sticky fluid, which can break up into smaller droplets or merge to create larger ones. The initial drops are planet-sized; the droplets are smaller, but only in comparison to planets. They're still fairly big.

A standard model for fluid dynamics goes back to Leonhard Euler and Daniel Bernoulli in the 1700s. It formulates the physical laws of fluid flow as a partial differential equation, which describes how the velocity of the fluid at each point in space changes over time in response to the forces that act. Except in very simple cases, it's not possible to find formulas that solve the equation, but very accurate computational methods have been devised. A major issue is the nature of the model, which in principle requires us to study the fluid velocity at every point in some region of space. However, computers can't do infinitely many calculations, so we 'discretise' the equation: approximate it by a related equation involving only a finite number of points. The simplest method uses the points of a grid as a representative sample of the entire fluid, and keeps track of how the velocity changes at the grid points. This approximation is good if the grid is fine enough.

Unfortunately this approach isn't great for colliding drops, because the velocity field becomes discontinuous when the drops break up. A cunning variant of the grid method comes to the rescue. It works even when droplets fragment or merge. This method, called smoothed particle hydrodynamics, breaks the fluid up into neighbouring 'particles' – tiny regions. But instead of using a fixed grid, we follow the particles as they respond to the forces that act. If nearby particles move with much the same speed and direction, they're in the same droplet and are going to stay in that droplet. But if neighbouring particles head off in radically different directions, or have significantly different velocities, the droplet is breaking up.

The mathematics makes this work by 'smoothing' each particle into a sort of soft fuzzy ball (jargon: spherical overlapping kernel function) and superposing these balls. The motion of the fluid is calculated by combining the motions of the fuzzy balls. Each ball can be represented

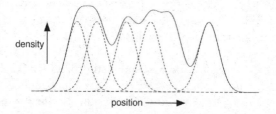

Representing the density of a fluid (solid line) as a sum of
small fuzzy droplets (dashed bell-shaped curves).

by its central point, and we have to calculate how these points move as time passes. Mathematicians call this kind of equation an n-body problem, where n is the number of points, or, equivalently, the number of fuzzy balls.

✦

All very well, but n-body problems are hard. Kepler studied a two-body problem, the orbit of Mars, and deduced it's an ellipse. Newton proved mathematically that when two bodies move under inverse square law gravity, both of them orbit in ellipses about their common centre of mass. But when mathematicians of the eighteenth and nineteenth centuries tried to understand the three-body problem – Sun, Earth, Moon is the basic case – they discovered it's nowhere near as neat and tidy. Even Delaunay's mammoth formula is only an approximation. In fact, the orbits are typically chaotic – highly irregular – and there are no nice formulas or classical geometric curves to describe them. See Chapter 9 for more about chaos.

To model a planetary collision realistically, the number n of fuzzy balls must be large – a thousand, or better still a million. Computers can calculate with big numbers, but here n doesn't characterise the numbers that appear in the sums: it measures how *complicated* the sums are. Now we run into the 'curse of dimensionality', where the dimension of a system is how many numbers you need to describe it.

Suppose we use a million balls. It takes six numbers to determine the state of each ball: three for its coordinates in space, three more for the components of its velocity. That's 6 million numbers, just to define the state at any instant. We want to apply the laws of mechanics and gravity to predict the future motion. These laws are differential equations, which determine the state a tiny period of time into the future, given the state now. Provided the time step into the future is very small, a second perhaps, the result will be very close to the correct future state. So now we're doing a sum with 6 million numbers. More precisely, we're doing 6 *million sums* with 6 million numbers: one sum for each number required for the future state. So the complexity of the calculation is 6 million *multiplied by* 6 million. That's 36 trillion. And this calculation tells us only what the next state is, one second into the future. Do the same again and we find out what happens two seconds from now, and so on. To find out what happens a thousand years ahead, we're looking at a period of about 30 billion seconds, and the complexity of the calculation is 30 billion times 36 trillion – about 10^{24}, one septillion.

And that's not the worst of it. Although each individual step may be a good approximation, there are now so many steps that even the tiniest error might grow, and big calculations take a lot of time. If the computer could do one step per second – that is, work in 'real time' – the sums would take a thousand years. Only a supercomputer could even come close to that. The only way out is to find a cleverer way to do the sums. In the early stages of the impact, a time step as short as a second might be needed because everything is a complicated mess. Later, a longer time step might be acceptable. Moreover, once two points get far enough apart, the force between them is so small that it might be possible to neglect it altogether. Finally – and this is where the main improvement comes from – the entire calculation might be simplified by setting it up in a more cunning way.

The earliest simulations made a further simplification. Instead of doing the calculations for three-dimensional space, they cut the problem down to two dimensions by assuming everything happens in the plane of the Earth's orbit. Now two circular bodies collide instead of two spherical ones. This simplification offers two advantages. That 6 million becomes only 4 million (four numbers per fuzzy ball). Better still, you no longer need a million balls; maybe 10,000 are enough. Now you have 40,000 in place of 6 million, and the complexity reduces from 36 trillion to 1·6 billion.

Oh, and one more thing…

It's not enough to do the calculation once. We don't know the mass of the impactor, its speed, or the direction it's coming from when it hits. Each choice requires a new calculation. This was a particular limitation of early work, because computers were slower. Time on a supercomputer was expensive, too, so research grants allowed only a small number of runs. Consequently the investigator had to make some good guesses, right at the start, based on rule-of-thumb considerations such as 'can this assumption give the right size for the final angular momentum?' And then hope.

The pioneers overcame these obstacles. They found a scenario that worked. Later work refined it. The origin of the Moon had been solved.

✦

Or had it?

Simulating the giant impact theory of the Moon's formation involves two main phases: the collision itself, creating a disc of debris, and the subsequent accretion of part of this disc to form a compact lump, the nascent Moon. Until 1996 researchers confined their calculations to the first phase, and their main method was smoothed particle hydrodynamics. Robin Canup and Erik Asphaug, writing in 2001, stated[5] that this method 'is well suited to intensely deforming systems evolving within mostly empty space,' which is exactly what we want for this phase of the problem.

Because these simulations are big and difficult, investigators contented themselves with working out what happened immediately after the impact. The results depend on many factors: the mass and speed of the impactor, the angle at which it hits the Earth, and the rotational speed of the Earth, which, several billion years ago, might well have been different from what it is today. The practical limitations of n-body computations meant that, to begin with, many alternatives were not explored. To keep the computations within bounds, the first models were two-dimensional. Then it was a matter of seeking cases where the impactor kicked a lot of material from the Earth's mantle into space. The most convincing example involved a Mars-sized impactor, so this became the prime contender.

All of these giant impact simulations had one feature in common: the impact created a huge disc of debris orbiting the Earth. The simulations usually modelled the dynamics of this disc for only a few orbits, enough to show that plenty of debris stayed in orbit rather than crashing back down again or heading off into outer space. It was *assumed* that many of the particles in the debris disc would eventually aggregate to form a large body, and that body would become the Moon, but no one checked this assumption because tracking the particles further would be too expensive and time-consuming.

Some of the later work made the tacit assumption that the main parameters – mass of impactor and so on – had already been sorted out by this pioneering work, and concentrated on calculating extra detail, rather than looking at alternative parameters. The pioneering work became a kind of orthodoxy, and some of its assumptions ceased to be questioned. The first sign of trouble came early on. The only scenarios that gave a plausible fit to observations required the impactor to graze the Earth rather than smash into it head on, so the impactor could not have been in the Earth's orbital plane. The two-dimensional model is inadequate, and only a full three-dimensional simulation can do the job. Fortunately the powers of supercomputers evolve rapidly, and with enough time and expense it became possible to analyse collisions in three-dimensional models.

However, most of these improved simulations showed that the Moon should contain a lot of rock from the *impactor*, and much less from the Earth's mantle. So the original simple explanation of the similarity between Moon rock and the mantle became much less convincing: it seemed to require Theia's mantle to be amazingly similar to that of the Earth. Some astronomers nevertheless maintained that this was what must have happened, neatly forgetting that such a similarity between Earth and Moon was one of the puzzles that the theory was supposed to explain. If it didn't wash for the Moon, why was it acceptable for Theia?

There's a partial answer: maybe Theia and the Earth originally formed at about the same distance from the Sun. The objections raised earlier for the Moon don't apply. There's no issue with angular momentum, because we haven't a clue what the other chunks of Theia did after the impact. And it's reasonable to assume that bodies that formed at similar locations in the solar nebula have similar compositions. But it's still hard to explain why Earth and Theia stayed separate for long enough to become planets in their own right – but then collided. It's not impossible, but it doesn't look likely.

A different theory seems more plausible, because it makes no assumptions about Theia's composition. Suppose that after the silicate rocks vaporised, and before they started to aggregate, they were thoroughly mixed. Then both Earth and Moon would have received donations of very similar rock. Calculations indicate that this idea works only if the vapour stays around for about a century, forming a kind of shared atmosphere spread along the common orbit of Theia and the Earth. Mathematical studies are under way to decide whether this theory is dynamically feasible.

Be that as it may, the original idea that the impactor splashed off a chunk of Earth's mantle, but did not itself contribute much to the eventual Moon, would be much more convincing. So astronomers sought alternatives, still involving a collision, but based on very different assumptions. In 2012 Andreas Reufer and coworkers analysed the effects of a fast-moving impactor much larger than Mars that sideswipes the Earth instead of colliding head-on.[6] Very little of the material splashed off comes from the impactor, the angular momentum works out fine, and the composition of the mantle and the Moon are even more similar than previously thought. According to a new analysis of *Apollo* lunar rock by Junjun Zhang's team, the ratio of titanium-50 and titanium-47 isotopes is the same as for the Earth within four parts per million.[7]

Other possibilities have been studied as well. Matja Cuk and coworkers have shown that the correct chemistry of Moon rocks and total angular momentum could have arisen from a collision with a smaller impactor, provided the Earth was spinning much faster than it is today. The spin changes the amount of rock that splashes off, and which body it comes from. After the collision, gravitational forces from the Sun and Moon could have slowed the Earth's spin. On the other hand, Canup has found convincing simulations in which the Earth was spinning only marginally faster than today, by assuming the impactor was significantly bigger than Mars. Or perhaps two bodies, five times the size of Mars collided, then recollided, creating a large disc of debris that eventually formed the Earth and Moon. Or...

✦

Or possibly the original impactor theory is correct, Theia *did* have much the same composition as the Earth, and that wasn't a coincidence at all.

In 2004 Canup[8] showed that the most plausible type of Theia should have about one-sixth of the mass of the Earth, and that four-fifths of the resulting Moon's material should have come from Theia. This implies that Theia's chemical composition must have been as close to that of the Earth as the Moon's is. This seems very unlikely: the bodies of the solar system differ considerably from each other, so what was different about Theia? As we've seen, a possible answer is that Earth and Theia formed under similar conditions – at a similar distance from the Sun, so they both swept up the same stuff. Moreover, being in roughly the same orbit improves the chance of a collision.

On the other hand, could two large bodies form in the same orbit? Wouldn't one of them win by sweeping up most of the available material? You can argue about this forever ... or you can do the sums. In 2105 Alessandra Mastrobuono-Battisti and coworkers used n-body methods to make 40 simulations of the late stages of planetary accretion.[9] By then, Jupiter and Saturn are fully formed, they've sucked up most of the gas and dust, and planetesimals and larger 'planetary embryos' are coming together to form the really large bodies. Each run started with about 85–90 planetary embryos and 1000–2000 planetesimals, lying in a disc between 0·5 and 4·5 AU. The orbits of Jupiter and Saturn were inclined slightly to this disc, and the inclinations differed between runs.

In most runs, about three or four rocky inner planets formed within 100–200 million years, as the embryos and planetesimals merged. The simulation kept track of each world's feeding zone, the region from which its components were gobbled up. On the assumption that the chemistry of the solar disc depends mainly on distance from the Sun, so that bodies in equidistant orbits have much the same composition, we can compare the chemical compositions of impacting bodies. The team focused on how each of the three or four surviving planets compares to its most recent impactor. Tracking back through these bodies' feeding zones leads to probability distributions for the composition of each body. Then statistical methods determine how similar these distributions are. The impactor and the planet have much the same composition in about one-sixth of the simulations. Taking into account the likelihood that some of the proto-planet also gets mixed into the Moon, this figure doubles to about one-third. In short: there is about *one chance in three* that Theia would have had the same chemistry as Earth. This is entirely plausible, so despite previous concerns, the similar chemistry of Earth's mantle and the Moon's surface rock is, in fact, consistent with the original giant impact scenario.

Right now, we have an embarrassment of riches: several distinct giant impact theories, all in good agreement with the main evidence. Which, if any, is correct remains to be seen. But to get both the chemistry and the angular momentum right, a large impactor seems unavoidable.

The Clockwork Cosmos

But should the Lord Architect have left that space empty? Not at all.

Johann Titius, in Charles Bonnet's *Contemplation de la Nature*

NEWTON'S *PRINCIPIA* ESTABLISHED the value of mathematics as a way to understand the cosmos. It led to the compelling notion of a clockwork universe, in which the Sun and planets were created in their present configuration. The planets went round and round the Sun in roughly circular orbits, nicely spaced so that they didn't run into each other – didn't even come close. Although everything jiggled about a bit, thanks to each planet's gravity tugging every other planet, nothing important changed. This view was encapsulated in a delightful gadget known as an orrery: a desktop machine in which tiny planets on sticks moved round and round the central Sun, driven by cogwheels. Nature was a gigantic orrery, with gravity for gears.

Mathematically minded astronomers knew that it wasn't quite that simple. The orbits weren't exact circles, they didn't even lie in the same plane, and some of the jiggles were quite substantial. In particular the two largest planets in the solar system, Jupiter and Saturn, were engaged in some kind of long-term gravitational tug-of-war, pulling each other first ahead of their usual positions in their orbits, then behind, over and over again. Laplace explained this around 1785. The two giants are close to a 5:2 resonance, in which Jupiter goes round the Sun five times while Saturn goes round twice. Measuring their positions in orbit as angles, the difference

2 × angle for Jupiter − 5 × angle for Saturn

is close to zero – but, as Laplace explained, it's not exactly zero. Instead, it slowly changes, completing a full circle every 900 years. This effect became known as the 'great inequality'.

Laplace proved that the interaction doesn't produce large changes to the eccentricity or inclination of either planet's orbit. This kind of result led to a general feeling that the current arrangement of the planets is stable. It would be much the same far into the future, and it had always been that way in the past.

Not so. The more we learn about the solar system, the less it looks like clockwork, and the more it looks like some bizarre structure that, although *mostly* well behaved, occasionally goes completely crazy. Remarkably, these weird gyrations don't cast doubt on Newton's law of gravity: they are *consequences* of it. The law itself is mathematically neat and tidy, simplicity itself. But what it leads to is not.

✦

To understand the origins of the solar system, we must explain how it arose and how its multifarious bodies are arranged. At first sight they're a pretty eclectic lot – each world is a one-off, and the differences outweigh the similarities. Mercury is a hot rock that revolves three times every two orbits, a 3:2 spin–orbit resonance. Venus is an acid hell whose entire surface reformed a few hundred million years ago. Earth has oceans, oxygen, and life. Mars is a frigid desert with craters and canyons. Jupiter is a giant ball of coloured gases making decorative stripes. Saturn is similar, though less dramatic, but in compensation it has gorgeous rings. Uranus is a docile ice giant and it spins the wrong way. Neptune is another ice giant, with encircling winds that exceed 2000 kilometres per hour.

However, there are also tantalising hints of order. The orbital distances of the six classical planets, in astronomical units, are:

Mercury	0·39
Venus	0·72
Earth	1·00
Mars	1·52
Jupiter	5·20
Saturn	9·54

The numbers are a bit irregular, and at first it's hard to find a pattern, even if it occurs to you to look. But in 1766 Johann Titius spotted something interesting in these numbers, and described it in his translation of Charles Bonnet's *Contemplation de la Nature*:

> Divide the distance from the Sun to Saturn into 100 parts; then Mercury is separated by four such parts from the Sun, Venus by 4+3 = 7 such parts, the Earth by 4+6 = 10, Mars by 4+12 = 16. But notice that from Mars to Jupiter there comes a deviation from this so exact progression. From Mars there follows a space of 4+24 = 28 such parts, but so far no planet was sighted there. ... Next to this for us still unexplored space there rises Jupiter's sphere of influence at 4+48 = 52 parts; and that of Saturn at 4+96 = 100 parts.

Johann Bode mentioned the same numerical pattern in 1772 in his *Anleitung zur Kenntniss des Gestirnten Himmels* (Manual for Knowing the Starry Sky), and in later editions he credited it to Titius. Despite that, it's often called Bode's law. A better term, now in general use, is Titius–Bode law.

This rule, which is purely empirical, relates planetary distances to a (nearly) geometric sequence. Its original form started with the sequence 0, 3, 6, 12, 24, 48, 96, 192, in which each number after the second is twice its predecessor, and added 4 to them all, getting: 4, 7, 10, 16, 28, 52, 100. However, it's useful to bring these numbers into line with current units of measurement (AU) by dividing them all by ten, giving: 0·4, 0·7, 1·0, 1·6, 2·8, 5·2, 10·0. These numbers fit the spacing of the planets surprisingly well, except for a gap corresponding to 2·8. Titius thought he knew what must be in that gap. The portion of his remark that I replaced by an ellipsis (...) reads:

> But should the Lord Architect have left that space empty? Not at all. Let us therefore assume that this space without doubt belongs to the still undiscovered satellites of Mars, let us also add that perhaps Jupiter still has around itself some smaller ones which have not been sighted yet by any telescope.

We now realise that the satellites of Mars will be found close to Mars, and ditto for Jupiter, so Titius was a bit shy of the mark in some respects, but the proposal that *some* body ought to occupy the gap was spot on. However, no one took it seriously until Uranus was discovered in 1781, and it also fitted the pattern. The predicted distance is 19·6; the actual one is 19·2.

Encouraged by this success, astronomers started looking for a previously unobserved planet circling the sun at about 2·8 times the radius of Earth's orbit. In 1801 Giuseppe Piazzi found one – ironically, just before a systematic search got under way. It was given the name Ceres, and we take up its story in Chapter 5. It was smaller than Mars, and much smaller than Jupiter, but it was *there*.

To make up for its diminutive stature, it was not alone. Three more bodies – Pallas, Juno, and Vesta – were soon found at similar distances. These were the first four asteroids, or minor planets, and they were soon followed by many more. About 200 of them are more than a kilometre across, over 150 million at least 100 metres across are known, and there are expected to be millions that are even smaller. They famously form the asteroid belt, a flat ring-shaped region between the orbits of Mars and Jupiter.

Other small bodies exist elsewhere in the solar system, but the first few discoveries added weight to Bode's view that the planets are distributed in a regular manner. The subsequent discovery of Neptune was motivated by discrepancies in the orbit of Uranus, not by the Titius–Bode law. But the law predicted a distance of 38·8, reasonably close to the actual distance, between 29·8 and 30·3. The fit is poorer but acceptable. Then came Pluto: theoretical distance 77·2, actual distance between 29·7 and 48·9. Finally, the Titius–Bode 'law' had broken down.

Other typical features of planetary orbits had also broken down. Pluto is very strange. Its orbit is highly eccentric and tilted a whopping 17 degrees away from the ecliptic. Sometimes Pluto even comes *inside* the orbit of Neptune. Unusual features like this recently led to Pluto being reclassified as a dwarf planet. In partial compensation, Ceres also became a dwarf planet, not a mere asteroid (or minor planet).

Despite its mix of success and failure, the Titius–Bode law poses an important question. Is there some mathematical rationale to the spacing of the planets? Or could they, in principle, have been spaced in any desired manner? Is the law a coincidence, a sign of an underlying pattern, or a bit of both?

✦

The first step is to reformulate the Titius–Bode law in a more general and slightly modified way. Its original form has an anomaly: the use of 0 as the first term. To get a geometric sequence this ought to be 1·5. Although this choice makes the distance for Mercury 0·55, which is less accurate, the whole game is empirical and approximate, so it makes more sense to keep the mathematics tidy and use 1·5. Now we can express the law in a simple formula: the distance from the Sun to the nth planet, in astronomical units, is

$$d = 0{\cdot}075 \times 2^n + 0{\cdot}4$$

Now we must do a few sums. In the grand scheme of things, 0·4 AU doesn't make much difference for the more distant planets, so we remove it to get $d = 0{\cdot}075 \times 2^n$. This is an example of a power law formula, which in general looks like $d = ab^n$, where a and b are constants.

Take logarithms:

$$\log d = \log a + n \log b$$

Using n and $\log d$ as coordinates, this is a straight line with slope $\log b$, meeting the vertical axis at $\log a$. So the way to spot a power law is to perform a 'log/log plot' of $\log d$ against n. If the result is close to a straight line, we're in business. Indeed, we can do this for quantities other than the distance d, for example the period of revolution around the star or the mass.

If we try this for the distances of planets, including Ceres and Pluto, we obtain the left-hand picture. Not far from a straight line, as we'd expect from the Titius–Bode law. What about the masses, shown in the right-hand picture? This time the log/log plot is very different. No sign of a straight line – or of any clear pattern.

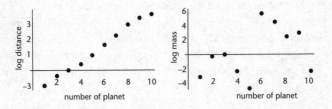

Left: Log/log plot of planetary distances lies close to a straight line.
Right: Log/log plot of planetary masses doesn't look like a straight line.

The orbital periods? A nice straight line again: see the left-hand picture. However, that's no surprise, because Kepler's third law relates to period to the distance in a manner that preserves power law relationships. Searching further afield, we examine the five main moons of Uranus, and get the right-hand picture. Power law again.

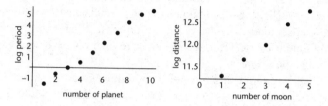

Left: Log/log plot of planetary periods lies close to a straight line.
Right: Log/log plot of distances of moons of Uranus lies close to a straight line.

✦

Coincidence, or something deeper? Astronomers are divided. At best there seems to be a *tendency* towards power law spacings. They're not universal.

There may be a rational explanation. The most likely one starts from the idea that the dynamics of a random system of planets depends crucially on resonances: cases where two planets have orbital periods in some simple fractional relationship. For example, one period might be 3/5 of the other, a 5:3 resonance.[1] Ignoring all other bodies, these two planets will keep lining up with each other, along the radial line from the star, at regular intervals, because five rotations of one match three of the other perfectly. Over long periods of time the resulting small disturbances will accumulate, so the planets will tend to change their orbits. For period ratios that are not simple fractions, on the other hand, the disturbances tend to cancel out, because there's no dominant direction for the force of gravity acting between the two worlds.

This isn't just a vague suggestion: detailed calculations and an extensive body of mathematical theory support it. To a first approximation, the orbit of a celestial body is an ellipse. At the next level of approximation, the ellipse precesses: its major axis slowly rotates.

Approximating even more accurately, the dominant terms in formulas for the motion of celestial bodies come from secular resonances – more general types of resonant relation among the periods with which the orbits of several bodies precess.

Precisely how resonant bodies move depends on the ratio of their periods, as well as their locations and velocities, but often the result is to clear out such orbits. Computer simulations indicate that randomly spaced planets tend to evolve into positions that satisfy relationships roughly similar to the Titius–Bode law, as resonances sweep out gaps. But it's all a bit vague.

The solar system contains several 'miniature' systems, namely, the moons of the giant planets. Jupiter's three largest satellites, Io, Europa, and Ganymede, have orbital periods in ratios close to 1:2:4, each twice the previous one (see Chapter 7). The fourth, Callisto, has a period slightly less than twice that of Ganymede. By Kepler's third law, the orbital radii are similarly related, except that the multiplier 2 has to be replaced by its 2/3 power, which is 1·58. That is, the orbital radius of each satellite is roughly 1·58 times that of the previous one. This is a case where resonance stabilises orbits instead of clearing them out, and the ratio of distances is 1·58 rather than the 2 of the Titius–Bode law. But the spacings still satisfy a power law. The same goes for the moons of Saturn and Uranus, as Stanley Dermott pointed out in the 1960s.[2] Such a spacing is called 'Dermott's law'.

Power law spacings are a more general pattern that includes a good approximation to the Titius–Bode law. In 1994 Bérengère Dubrulle and François Graner derived power law spacings for typical collapsing solar nebulas[3] by applying two general principles. Both depend on symmetry. The cloud is axially symmetric, and the matter distribution is much the same on all scales of measurement, a symmetry of scale. Axial symmetry makes dynamic sense because an asymmetric cloud will either break up or become more symmetric as time passes. Scale symmetry is typical of important processes believed to influence planet formation, such as turbulent flow within the solar nebula.

Nowadays we can look beyond the solar system. All hell breaks loose: the orbits of known exoplanets – planets round other stars – have all sorts of spacings, most of them very different from what we find in the solar system. On the other hand, the known exoplanets are an imperfect sample of those that actually exist; often only one planet is known for a given star, even though it probably has others. The detection methods are biased towards finding large planets circling close to their primaries.

Until we can map out the *entire* planetary systems of many stars, we won't really know what exoplanetary systems look like. But in 2013 Timothy Bovaird and Charles Lineweaver looked at 69 exoplanet systems known to have at least four planets, and 66 of them obey power laws. They also used the resulting power laws to tentatively predict 'missing' planets – doing a Ceres on an exosystem. Of the 97 planets predicted in this manner, only five have so far been observed. Even allowing for the difficulty of detecting small planets, this is a bit disappointing.

All of this is rather tentative, so attention has shifted to other principles that might explain how planetary systems are organised. These rely on subtle details of nonlinear dynamics and are not merely empirical. However, the patterns are less obviously numerical. In particular, Michael Dellnitz has shown mathematically that Jupiter's gravitational field seems to have arranged all of the other planets into an interconnected system linked by a natural set of 'tubes'. These tubes, which can be detected only through their mathematical features, provide natural low-energy routes between the different worlds. We'll discuss this idea along with related matters in Chapter 10, where it fits more naturally.

✦

Coincidence or not, the Titius–Bode law inspired some important discoveries.

The only planets visible to the naked eye are the classical five: Mercury, Venus, Mars, Jupiter, and Saturn. Plus Earth, if you want to be pedantic, but we only ever see a small part of it at one time. With the invention of the telescope, astronomers could observe stars that are too dim to see with the eye alone, along with other objects such as comets, nebulas, and satellites. Working at the limits of what was then technically feasible, the early astronomers often found it easier to spot a new object than to decide what it was.

Exactly this problem confronted William Herschel in 1781, when he pointed the telescope in the garden of his house in Bath towards the constellation Taurus and noticed a faint spot of light near the star zeta Tauri, which at first he thought was either 'a Nebulous Star or perhaps a Comet'. Four nights later he wrote in his journal that he had 'found it was a Comet, for it has changed its place'. About five weeks after that, when he reported his discovery to the Royal Society, he still described it as a comet. If you observe a star using lenses which magnify by different amounts, it remains pointlike even at the highest magnification, but this new object seemed to become larger as the magnification increased – 'as planets are', he remarked. But the same goes for comets, and Herschel was convinced he'd discovered a new comet.

As more information came in, some astronomers begged to differ, among them the Astronomer Royal Nevil Maskelyne, Anders Lexell, and Bode. By 1783 there was a consensus that the new object was a planet, and it required a name. King George III had given Herschel £200 a year on condition that he moved close enough to Windsor Castle for the royal family to look through his telescopes. Herschel, minded to repay him, wanted to call it Georgium Sidus, 'George's Star'. Bode suggested Uranus, the Latin form of Ouranos, the Greek sky god, and this name won the day, despite being the only planetary name based on a Greek god rather than a Roman one.

Laplace, quick off the mark, calculated the orbit of Uranus in 1783. The period is 84 years and the average distance from the Sun is about 19 AU or 3 billion kilometres. Although almost circular, Uranus's orbit is more eccentric than that of any other known planet, with a radius that ranges from 18 to 20 AU. Over the years, better telescopes made it possible to measure the planet's rotational period, which is 17 hours 14 minutes, and to reveal that it's retrograde – the planet spins in the opposite direction to every other. Its axis is tilted through more than a right angle, lying pretty much in the ecliptic plane of the solar system instead of being roughly perpendicular to it. As a result, Uranus experiences an extreme form of midnight sun: each pole endures 42 years of daylight followed by 42 of darkness, with one pole being dark while the other is light.

Clearly there's something strange about Uranus. On the other hand, it fits the Titius–Bode law perfectly.

Once the orbit was known and past sightings could be associated with the new world, it became apparent that it had been spotted before, but misidentified as a star or a comet. Indeed, it's just visible with good eyesight, and it was plausibly one of the 'stars' in Hipparchus's catalogue of 128 BC, and, later, in Ptolemy's *Almagest*. John Flamsteed observed it six times in 1690, thinking it was a star, then named 34 Tauri. Pierre Lemonnier observed it twelve times between 1750 and 1769. Although Uranus is a planet, it moves so slowly that it's easy not to notice any change in its position.

✦

So far, the main role of mathematics in understanding the solar system had been mainly descriptive, reducing lengthy series of observations to a simple elliptical orbit. The only prediction emerging from the mathematics was to forecast the planet's position in the sky at future dates. But, as time passed and enough observations accumulated, Uranus increasingly appeared to be in the wrong place. Alexis Bouvard, a student of Laplace, made numerous high-precision observations of Jupiter, Saturn, and Uranus, as well as discovering eight comets. His tables of the motion of Jupiter and Saturn proved to be very accurate, but Uranus drifted steadily away from its predicted location. Bouvard suggested that an even more distant planet might be perturbing Uranus's orbit.

'Perturb' here means 'have an effect on'. If we could express that effect mathematically in terms of the orbit of this presumptive new planet, we could work backwards to deduce that orbit. Then astronomers would know where to look, and if the prediction was based on fact, they could find the new planet. The big snag with this approach is that the motion of Uranus is influenced significantly by the Sun, Jupiter, and Saturn. The rest of the solar system's bodies can perhaps be neglected, but we still have five bodies to deal with. No exact formulas are known for three bodies; five are much harder.

Fortunately, the mathematicians of the day had already thought of a clever way to get round that issue. Mathematically, a perturbation of a system is a new effect that changes the solutions of its equations. For example, the movement of a pendulum under gravity *in a vacuum* has an elegant solution: the pendulum repeats the same oscillations over and again forever. If there's air resistance, however, the equation of motion changes to include this extra resistive force. This is a perturbation of the pendulum model, and it destroys the periodic oscillations. Instead, they die down and eventually the pendulum stops.

Perturbations lead to more complex equations, which are usually harder to solve. But sometimes the perturbation itself can be used to find out how solutions change. To do this we write down equations for the *difference* between the unperturbed solution and the perturbed one. If the perturbation is small, we can derive formulas that approximate this difference by neglecting terms in the equations that are much smaller than the perturbation. This trick simplifies the equations enough to solve them explicitly. The resulting solution is not exact, but it's often good enough for practical purposes.

If Uranus were the only planet, its orbit would be a perfect ellipse. However, this ideal orbit is perturbed by Jupiter, Saturn, and any other bodies of the solar system that we know about. Their combined gravitational fields change the orbit of Uranus, and this change can be described as a slow variation in the orbital elements of Uranus's ellipse. To a good approximation, Uranus always moves along *some* ellipse, but it's no longer always the *same* ellipse. The perturbations slowly change its shape and inclination.

In this manner we can calculate how Uranus would move when all important perturbing bodies are accounted for. The observations show that Uranus does not, in fact, follow this predicted orbit. Instead, it gradually deviates in ways that can be measured. So we add a hypothetical perturbation by an unknown Planet X, calculate the new perturbed orbit, set that equal to the observed one, and deduce the orbital elements of Planet X.

In 1843, in a computational *tour de force*, John Adams calculated the hypothetical new world's orbital elements. By 1845 Urbain Le Verrier was carrying out similar calculations independently. Adams sent his predictions to George Airy, the current British Astronomer Royal, asking him to look for the predicted planet. Airy was worried about some aspects of the calculation – wrongly, as it transpired – but Adams was unable to reassure him, so nothing was done. In 1846 Le Verrier published his own prediction, which again aroused little interest, until Airy noticed that both mathematicians had come up with very similar results. He instructed James Challis, Director of the Cambridge Observatory, to look for the new planet, but Challis failed to find anything.

Soon after, however, Johann Galle spotted a faint point of light about 1 degree away from Le Verrier's prediction and 12 degrees from Adams's. Later, Challis found that he'd observed the new planet twice, but didn't have an up-to-date star map and had generally been a bit sloppy, so he'd missed it. Galle's spot of light was another new planet, later named Neptune. Its discovery was a major triumph of celestial mechanics. Now mathematics was revealing the existence of unknown worlds, not just codifying the orbits of known ones.

✦

The solar system now boasted eight planets and a rapidly growing number of 'minor planets', or asteroids (see Chapter 5). But even prior to the discovery of Neptune, some astronomers, among them Bouvard and Peter Hansen, were convinced that a single new body couldn't explain the anomalies in the motion of Uranus. Instead, they believed that the discrepancies were evidence for *two* new planets. This idea was batted to and fro for another 90 years.

Percival Lowell founded an observatory in Flagstaff Arizona in 1894, and 12 years later he decided to sort out once and for all the anomalies in the orbit of Uranus, starting a project that he named Planet X. Here X is the mathematical unknown, not a Roman numeral (which would have been IX anyway). Lowell had somewhat ruined his scientific reputation by promoting the idea of 'canals' on Mars, and wanted to restore it: a new planet would be ideal. He used mathematical methods to predict where this hypothetical world should be, and then made a systematic search – with no result. He tried again in 1914–16, but again found nothing.

Meanwhile Edward Pickering, Director of Harvard College Observatory, had come up with his own prediction: Planet O, at a distance of 52 AU. By then the British astronomer Philip Cowell had declared the whole search a wild goose chase: the supposed anomalies in the motion of Uranus could be accounted for by other means.

In 1916 Lowell died. A legal dispute between his widow and the observatory put paid to any further search for Planet X until 1925, when Lowell's brother George paid for a new telescope. Clyde Tombaugh was given the job of photographing regions of the night sky twice, two weeks apart. An optical device compared the two images, and anything that had changed position would blink, drawing attention to the movement. He took a third image to resolve any uncertainties. Early in 1930, he was examining an area in Gemini and something blinked. It was within 6 degrees of a location suggested by Lowell, whose prediction appeared to be upheld. Once the object was identified as a new planet, a search of the archives showed that it had been photographed in 1915 but not then recognised as a planet.

The new world was dubbed Pluto, its first two letters being Lowell's initials.

Pluto turned out to be much smaller than expected, with a mass only one tenth that of the Earth. That implied that it could not, in fact, explain the anomalies that had led Lowell and others to predict its existence. When the low mass was confirmed in 1978, a few astronomers resumed the search for Planet X, believing that Pluto was a red herring and a more massive unknown planet must be out there somewhere. When Myles Standish used data from the 1989 Voyager fly-by of Neptune to refine the figure for Neptune's mass, the anomalies in the orbit of Uranus vanished. Lowell's prediction was just a lucky coincidence.

Pluto is weird. Its orbit is inclined at 17 degrees to the ecliptic, and is so eccentric that for a time Pluto comes closer to the Sun than Neptune. However, there's no chance of them colliding, for two reasons. One is the angle between their orbital planes: their orbits cross only on the line where those planes meet. Even then, both worlds must pass through the same point on this line at the same time. This is where the second reason comes in. Pluto is locked into a 2:3 resonance with Neptune. The two bodies therefore repeat essentially the same movements every two orbits of Pluto and three of Neptune, that is, every 495 years. Since they haven't collided in the past, they won't do so in the future – at least, not until large-scale reorganisation of other bodies of the solar system disturbs their cosy relationship.

✦

Astronomers continued searching the outer solar system for new bodies. They discovered that Pluto has a comparatively large moon, Charon, but nothing else was spotted beyond the orbit of Neptune until 1992, when a small body christened (15760) 1992 QB_1 showed up. It was so obscure that this is still its name (a proposal to call it 'Smiley' was rejected because that name had already been used for an asteroid), but it proved to be the first of a gaggle of trans-Neptunian objects (TNOs), of which more than 1500 are known. Among them are a few larger bodies, though still smaller than Pluto: the largest is Eris, followed by Makemake, Haumea, and 2007 OR$_{10}$.

All of these objects are too lightweight and too distant to be predicted from their gravitational effects on other bodies, and were discovered by searching through images. But there are some noteworthy mathematical features, related to the effects of other bodies on *them*. Between 30 and 55 AU lies the Kuiper belt, most of whose members are in roughly circular orbits close the ecliptic. Some of these TNOs are in resonant orbits with Neptune. Those in 2:3 resonance are called plutinos, because they include Pluto. Those in 1:2 resonance – period twice that of Neptune – are called twotinos. The rest are classical Kuiper belt objects, or cubewanos;[4] they also have roughly circular orbits, but experience no significant perturbations from Neptune. Further out is the scattered disc. Here, asteroid-like bodies move in eccentric orbits, often inclined at a large angle to the ecliptic. Among them are Eris and Sedna.

As more and more TNOs were found, some astronomers began to feel that it made little sense to call Pluto a planet, but not Eris, which they thought was slightly bigger. Ironically, images from *New Horizons* showed that Eris is slightly smaller than Pluto.[5] But once other TNOs were classified as planets, some would be smaller than the asteroid (or minor planet) Ceres. After much heated debate, the International Astronomical Union demoted Pluto to the status of dwarf planet, where it was joined by Ceres, Haumea, Makemake, and Eris. New definitions of the terms 'planet' and 'dwarf planet' were carefully tailored to shoehorn the bodies concerned into those two classifications. However, it's not yet clear whether Haumea, Makemake, and Eris actually fit the definition. It's also suspected that a few hundred further dwarf planets exist in the Kuiper belt, and up to ten thousand in the scattered disc.

✦

When a new scientific trick works, it's only sensible to try it on similar problems. Used to predict the existence and location of Neptune, the perturbation trick worked brilliantly. Tried on Pluto, it seemed to work brilliantly as well, until astronomers realised that Pluto is too small to create the anomalies used to predict it.

The trick failed dismally for a planet named Vulcan. This is not the fictional planet of *Star Trek*, homeworld of Mr Spock, which according to the science fiction writer James Blish orbits the star 40 Eridani A. Instead, it's the fictional planet that orbits an obscure and rather ordinary star known to science fiction writers as Sol. Or, more familiarly, the Sun. Vulcan teaches us several lessons about science: not just the obvious one that mistakes can be made, but the more general point that awareness of past mistakes can stop us repeating them. Its prediction is linked to the introduction of relativity as an improvement on Newtonian physics. But more of that story later, as they say.

Neptune was discovered because of anomalies in the orbit of Uranus. Vulcan was proposed to explain anomalies in the orbit of Mercury – and the proposer was none other than Le Verrier, in work that predates Neptune. In 1840 the Director of the Paris Observatory, François Arago, wanted to apply Newtonian gravitation to the orbit of Mercury, and asked Le Verrier to perform the necessary calculations. When Mercury passed across the face of the Sun, an event called a transit, the theory could be tested by observing the times when the transit began and ended. There was a transit in 1843, and Le Verrier completed his calculations shortly beforehand, making it possible to predict the timing. To his dismay, the observations failed to agree with the theory. So Le Verrier went back to the drawing board, preparing a more accurate model based on numerous observations and 14 transits. And by 1859 he had noticed, and published, a small but baffling aspect of Mercury's motion that explained his original error.

The point at which Mercury's orbital ellipse gets nearest to the Sun, known as the perihelion (*peri* = close, *helios* = Sun), is a well-defined feature. As time passes, Mercury's perihelion slowly rotates relative to the background of distant ('fixed') stars. In effect, the entire orbit slowly pivots with the Sun at its focus; the technical term is precession. A mathematical result known as Newton's theorem of revolving orbits[6] predicts this effect as a consequence of perturbations by other planets. However, when Le Verrier plugged the observations into this theorem, the resulting numbers were very slightly wrong. Newtonian theory predicted that the perihelion of Mercury should precess through 532″ (seconds of arc) every hundred years; the observed figure was 575″. Something was causing an extra 43″ precession per century. Le Verrier suggested that some undiscovered planet, orbiting closer to the

Sun than Mercury, was responsible, and he named it Vulcan, after the Roman god of fire.

The Sun's glare would overwhelm any light reflected from such a closely orbiting planet, so the only practical way to observe Vulcan would be during a transit. Then it should be visible as a tiny dark dot. The amateur astronomer Edmond Lescarbault quickly announced that he'd found such a dot, which wasn't a sunspot because it moved at the wrong speed. Le Verrier announced the discovery of Vulcan in 1860, and on the strength of this he was awarded the Legion of Honour.

Unfortunately for Le Verrier and Lescarbault, a better-equipped astronomer, Emmanuel Liais, had also been observing the Sun at the behest of the Brazilian government, and had seen nothing of the kind. His reputation was at stake, and he denied that any such transit had occurred. The arguments became heated and confused. When Le Verrier died in 1877 he still believed he'd discovered another planet. Without Le Verrier's backing, the Vulcan theory lost momentum, and soon the consensus was straightforward: Lescarbault had been wrong. Le Verrier's prediction remained unverified, and there was widespread skepticism. Interest vanished almost totally in 1915, when Einstein used his new theory of general relativity to derive a precession of $42 \cdot 98''$ without any assumption of a new planet. Relativity was vindicated and Vulcan was thrown on the scrapheap.

We still don't know for certain that there are no bodies between Mercury and the Sun, though if one exists, it has to be very small. Henry Courten reanalysed images of the solar eclipse of 1970, stating that he'd detected at least seven such bodies. Their orbits couldn't be determined, and the claims haven't been confirmed. But the search for vulcanoids, as they're named, continues.[7]

5

Celestial Police

> The dinosaurs didn't have a space program, so they're not here
> to talk about this problem. We are, and we have the power to do
> something about it. I don't want to be the embarrassment of the
> galaxy, to have had the power to deflect an asteroid, and then not,
> and end up going extinct.
>
> Neil deGrasse Tyson, Space Chronicles

PURSUED BY A FLEET of interstellar warships firing sizzling bolts of
pure energy, a small band of courageous freedom fighters seeks refuge
in an asteroid belt, weaving violently through a blizzard of tumbling
rocks the size of Manhattan that constantly smash into each other.
The warships follow, evaporating the smaller rocks with laser beams
while accepting numerous hits from smaller fragments. In a cunning
manoeuvre, the fleeing vessel loops back on itself and dives into a deep
tunnel at the centre of a crater. But its worries have only begun...

It's a breathtaking cinematic image.

It's also total nonsense. Not the fleet of warships, the energy bolts,
or the galactic rebels. Not even the monstrous worm lurking at the end
of the tunnel. Those *might* just happen one day. It's that blizzard of
tumbling rock. No way.

I reckon it's all down to that badly chosen metaphor. Belt.

✦

Once upon a time the solar system, as then understood, lacked a belt.
Instead, there was a gap. According to the Titius–Bode law, there ought

to have been a planet between Mars and Jupiter, but there wasn't. If there had been, the ancients would have seen it and associated yet another god with it.

When Uranus was discovered, it fitted so neatly into the mathematical pattern of the Titius–Bode law that astronomers were encouraged to fill the gap between Mars and Jupiter. As we saw in the previous chapter, they succeeded. Baron Franz Xaver von Zach initiated the *Vereinigte Astronomische Gesellschaft* (United Astronomical Society) in 1800, with 25 members – among them Maskelyne, Charles Messier, William Herschel, and Heinrich Olbers. Because of its dedication to tidying up the unruly solar system, the group became known as the *Himmelspolizei* (celestial police). Each observer was assigned a 15-degree slice of the ecliptic and tasked with searching that region for the missing planet.

As is all too common in such matters, this systematic and organised approach was trumped by a lucky outsider: Giuseppe Piazzi, Astronomy Professor at the University of Palermo in Sicily. He wasn't searching for a planet; he was looking for a star, 'the 87th of the Catalogue of Mr. La Caille'. Early in 1801, near the star he was seeking, he saw another point of light, which matched nothing in the existing star catalogues. Continuing to observe this interloper, he found that it moved. His discovery was exactly where the Titius–Bode law required it to be. He named it Ceres after the Roman harvest goddess, who was also the patron goddess of Sicily. At first he thought he'd spotted a new comet, but it lacked the characteristic coma. 'It has occurred to me several times that it might be something better than a comet,' he wrote. Namely, a planet.

Ceres is rather small by planetary standards, and astronomers nearly lost it again. They had very little data on its orbit, and before they could obtain more measurements the motion of the Earth carried the line of sight to the new body too close to the Sun, so its faint light was swamped by the glare. It was expected to reappear a few months later, but the observations were so sparse that the likely position would be very uncertain. Not wishing to start the search all over again, astronomers asked the scientific community to provide a more reliable prediction. Carl Friedrich Gauss, then a relative unknown in the public eye, rose to the challenge. He invented a new way to deduce an orbit from three or more observations, now known as Gauss's method. When

Ceres duly reappeared within half a degree of the predicted position, Gauss's reputation as a great mathematician was sealed. In 1807 he was appointed Professor of Astronomy and Director of the Observatory at Göttingen University, where he remained for the rest of his life.

To predict where Ceres would reappear, Gauss invented several important numerical approximation techniques. Among them was a version of what we now call the fast Fourier transform, rediscovered in 1965 by James Cooley and John Tukey. Gauss's ideas on the topic were found among his unpublished papers and appeared posthumously in his collected works. He viewed this method as a form of trigonometric interpolation, inserting new data points between existing ones in a smooth manner. Today it's a vital algorithm in signal processing, used in medical scanners and digital cameras. Such is the power of mathematics, and what the physicist Eugene Wigner called its 'unreasonable effectiveness'.[1]

Building on this success, Gauss developed a comprehensive theory of the motion of small asteroids perturbed by large planets, which appeared in 1809 as *Theoria Motus Corporum Coelestium in Sectionibus Conicis Solem Ambientum* (Theory of Motion of Celestial Bodies Moving in Conic Sections around the Sun). In this work, Gauss refined and improved a statistical method introduced by Legendre in 1805, now called the method of least squares. He also stated that he'd had the idea first, in 1795, but (typically for Gauss) he hadn't published it. This method is used to deduce more accurate values from a series of measurements, each subject to random errors. In its simplest form it selects the value that minimises the total error. More elaborate variations are used to fit the best straight line to data about how one variable relates to another, or to deal with similar issues for many variables. Statisticians use such methods on a daily basis.

✦

When the orbital elements of Ceres were safely in the bag, so that it could be found whenever required, it turned out not to be alone. Similar bodies, of similar size or smaller, had similar orbits. The better your telescope, the more of them you could see, and the smaller they became.

Later in 1801 one of the Celestial Police, Olbers, spotted one such